A Guidebook to Nuclear Reactors

A Guidebook to

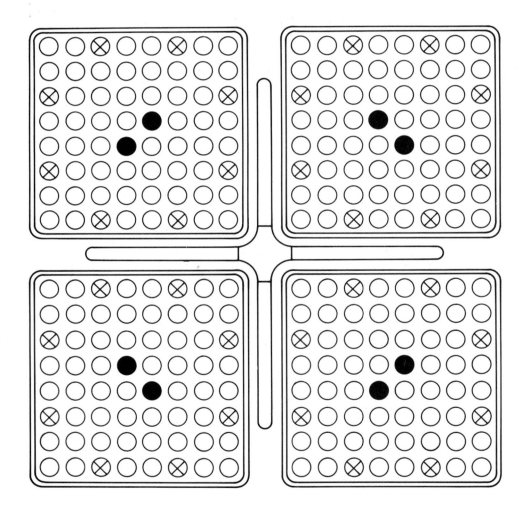

Nuclear Reactors

ANTHONY V. NERO, JR.

UNIVERSITY OF CALIFORNIA PRESS

BERKELEY / LOS ANGELES / LONDON

University of California Press
Berkeley and Los Angeles, California
University of California Press, Ltd.
London, England
Copyright © 1979 by
The Regents of the University of California
ISBN 0-520-03482-1
Library of Congress Catalog Card Number: 77-76183
Printed in the United States of America

1   2   3   4   5   6   7   8   9

Contents

Preface

THE PURPOSE of this book is to describe and illustrate reactors as they actually exist and to indicate the basic forms that nuclear systems may take in the future. The book also provides background information on many of the questions now facing nuclear power, including the issues of reactor safety and radioactive waste disposal, as well as the relative advantages of alternative fuel cycles: uranium-plutonium versus thorium-uranium. We hope that the discussion and description can provide a "feel" for the various reactor types, and for the issues of nuclear power, that is useful to individuals whose background includes no more than a single course in the physical sciences. At the same time, the information is sufficiently detailed to be of interest to many a professional interested in nuclear power.

There are, of course, a number of good treatments of the principles underlying reactor design, but they remain curiously isolated from specific designs and from commercial practice. Since nuclear reactors and fuel cycles draw the interest of many people, from aspiring nuclear engineers to active environmentalists, a publication emphasizing reactor systems, rather than their physics or engineering, seems useful. As it turns out, some of the physical principles are discussed here, but primarily to introduce the reader to basic aspects of nuclear reactors and to the features that distinguish one reactor type from another. We do not explain in detail how the particular parameters of each reactor are calculated, but do describe graphically the actual form of these reactors.

There are, on the other hand, several books that introduce the reader to the issues of nuclear power. These often argue strongly either for or against the use of nuclear reactors, yet present too little detailed information for a considered judgment. This book includes some detail on the questions about nuclear power: radioactive emissions, reactor safety, uranium utilization, reprocessing and waste disposal, and nuclear proliferation. For the most part, readers are left to draw their own conclusions. However, the book does argue that decisions on nuclear power ought to be made considering a variety of reactors, fuels, and fuel cycle configurations.

The discussion in the text is supported by numerous illustrations and tables. Many of these have been prepared specifically for this book, but most are drawn from other sources. About half are from publications of federal agencies, and we are grateful for their availability in the open literature. Many others have been provided kindly by reactor manufacturers, for which we thank them. In some cases, we have used prints prepared for Lawrence Berkeley Laboratory report LBL-5206, by the same title and author as this book; we appreciate having access to them.

This book has been prepared while the author was assistant professor of physics at Princeton University and, following that, physicist in the Energy and Environment Division of Lawrence Berkeley Laboratory. The author is grateful to individuals at these institutions and elsewhere who have encouraged this work. Special thanks are due to C. Blumstein, R. Budnitz, B. Casper, H. Feiveson, W. Loewenstein, A. Sesonske, F. von Hippel, and R. Whitesel, who generously made comments on parts of the manuscript at some stage of preparation. However, responsibility for any errors lies exclusively with the author.

November, 1977

Introduction

IN THE LAST decade, commercial nuclear power has emerged as one of the principal sources of energy in the United States and abroad. The electrical generating system has become an increasingly important component of the energy economy, and nuclear reactors have proved to be capable of generating the heat to drive the large turbine generators that supply a centralized electrical system. Although this system still depends primarily on fossil-fueled generating stations, a large share of new electrical generating capacity has been provided by nuclear power plants.

The fact that reactors depend directly on nuclear fission reactions, which leave highly radioactive materials as their residue, has generated another kind of heat as well. If inadequately contained, this residue can contaminate the biosphere and harm human beings. This fact, of itself, has led to substantial controversy over the adequacy of measures taken to limit such harm. In addition, the economic and even social cost of these measures has raised questions whether commercial nuclear power is really competitive with other methods of generating electricity, with other forms of energy, or with the cost of using less energy. And finally, the fact that the basic materials of commercial nuclear power can be fashioned into nuclear weapons may cause nuclear power to have a critical influence on world stability.

These questions are not easily answered. Man, and indeed the entire biosphere, has developed in a bath of natural background radiation that clearly exceeds the average exposure that he can ever expect to suffer as a result of commercial nuclear power operations. However, there will be some global increase in exposure due to nuclear power, and there is even some probability of local massive exposures due to large accidental releases of radioactivity from nuclear facilities. The social decision requires that the harm from these exposures be compared with the advantages that nuclear power may offer; such comparison of costs (monetary or not) and benefits should also be made for the alternatives to nuclear power, including the alternative of using less energy. In a similar way, the danger that widespread reliance on nuclear power may lead to wider access to nuclear weapons needs to be balanced against the perceived benefits of nuclear power. It may even be that, considering the other avenues to weapons production, shunning the use of nuclear power may decrease (rather than increase) stability, simply because of the precarious position some nations would be in were they to avoid using nuclear energy.

These questions are complicated by the fact that commercial nuclear power is not a monolithic institution. Nuclear electrical generation now depends almost entirely on one class of reactors, those that use water to carry heat from the reactor to the electrical generating system. These reactors now use uranium as their fuel and, until recently, planners had anticipated that plutonium produced in the course of reactor operation would be returned to reactors in fresh fuel, reducing the amount of uranium that is required. However, national and world resources of nuclear fuel are limited, and current water-cooled reactors, even those that recycle plutonium, deplete uranium resources at a substantial rate. Long-term and heavy reliance on nuclear electrical generation requires adoption of advanced reactor types that use nuclear fuels more efficiently; these could use either a uranium-plutonium fuel cycle or the potentially important thorium-uranium cycle. The alternative, of course, is to regard nuclear power as a short-term energy supply. Most nations do not have this point of view. This leads to the need to choose among the possible reactor types and to determine how these will be used, i.e., how their fuel will be supplied, how their wastes will be disposed of, how nuclear materials will be safeguarded, and so on.

The purpose of this book is to describe nuclear reactors and the context in which they are used. If it argues any point of view, it is only that more than one or two types of reactor exist and decisions taken without considering the large number of alternatives may be ill taken. The book describes the present status of nuclear power, emphasizing the various types of reactors that have been commercialized, and summarizes their health, safety, and resource-utilization characteristics. It then goes on to consider the manner in which nuclear power may develop and to describe the reactor systems that may form the basis of this development, emphasizing the differences between the uranium-plutonium and thorium-uranium fuel cycles.

The basic intention is to provide a framework within which choices about the course of nuclear power, and indeed about the existence of nuclear power, may be made. The style of the book is therefore more informational than argumentative. The emphasis is on the major features of reactors and on the major issues associated with their present and future use. The discussion is supported by many illustrations and tables, intended to convey a feeling for what nuclear reactors are and how they may be used.

The book is divided into four parts and is augmented by six appendixes which treat some technical aspects of nuclear reactors. Appendix A defines units and abbreviations used in the text. The reader is cautioned, though, that many of the numerical quantities cited in this book should be regarded as only representative; often the precise numbers change during redesign or reanalysis. Principal examples of such variations are reactor design parameters, which often change slightly from one reactor to another, even for the same type.

Part I of the text provides a general introduction to nuclear power plants, beginning with some of the basic design features of nuclear reactors, proceeding to the general question of material flows and environmental residuals in reactor systems, and then characterizing the size and significance of radioactive releases under both routine and accident conditions.

Part II describes reactors that have been commercially available, including light-water reactors, heavy-water reactors, and gas-cooled reactors. For each type, the basic systems, safety design, and operational characteristics are treated.

Part III discusses several questions that are basic to any future development of nuclear power: the extent of nuclear resources; the manner in which advanced reactors utilize these resources; the processing of nuclear fuels (including reprocessing and waste management operations); and the connection with nuclear weapons.

Finally, Part IV describes specific advanced reactors. These include breeder reactors, designed effectively to produce more nuclear fuel than they use, as well as modifications of current reactor types which can substantially increase the efficiency with which resources are utilized. Systems that combine fission reactors with other types of machines are also discussed briefly.

It appears that reliance on nuclear power will continue to increase, at least in the near future. However advantageous it may be to use energy more efficiently, the United States and other countries do not seem willing to set the limits to energy growth that would permit passing up the nuclear option, at least in this century. This situation may change, particularly in the United States, where varied energy supplies are available and where strong efforts to increase the effectiveness of energy use could limit the need for energy supplies. Nevertheless, it appears probable that nuclear power will remain, at some level, for a long period. Decisions on how much nuclear power should be used, and in what form, should include consideration of the various types of reactors and fuel cycles that are available and of the advantages and liabilities associated with their use. Providing such information is the purpose of this book.

Part I

A General Introduction to Nuclear Reactors

Nuclear power plants in the United States have recently reached sufficient number to produce a notable fraction of our total electrical power. A basic purpose of this book is to examine the types of reactors that may be employed for commercial power generation. However, before treating any specific designs, it is useful to introduce some basic considerations that underlie nuclear power plant design. We begin with four introductory chapters, which treat basic reactor design features, "input-output" characteristics of reactors, nuclear power plant emissions, and the potential for nuclear accidents. These chapters constitute an introduction to nuclear power plants and specify some concepts that will be useful in the discussion of particular reactor types and of possible directions for nuclear power.

Although a basic point of this book is that many types of reactors exist, with differing advantages and disadvantages, present-day nuclear electrical generation in the United States arises almost exclusively from one class of reactors. These employ ordinary water as their cooling fluid and as the material that slows down neutrons to increase their probability of causing the fissions from which energy is released. It is clear that, even as late as the year 2000, the bulk of nuclear electrical generation in the United States will utilize this type of reactor. Both for this reason and because of the greater information that is available on these reactors, portions of these introductory chapters will stress light-water reactors. This emphasis will be most obvious in the discussion of reactor safety. However, many of these introductory considerations are couched in terms that are applicable to other, although perhaps not all, reactor types. Instances where the discussion is not applicable will become clear in the later discussion of individual reactor types.

CHAPTER ONE

Basic Reactor Design Features

T HE PRIMARY societal and commercial interest in nuclear reactors arises from their potential for serving as heat sources for facilities that generate electrical power. Electricity can be generated in a number of ways, but the method by which the bulk of it is generated in the United States is to use thermal energy, heat, to produce steam, which drives a turbine-generator system. As in the accompanying illustration (Figure 1-1), this technique may be thought of as employing two basic systems: a steam supply system, which uses heat from the combustion of fossil fuels or from nuclear reactions to boil water, and an electrical generating system, which uses the resulting steam to produce electricity. In principle, even the sun may serve as the heat source for the steam supply system, but for the near future fossil-fueled boilers and nuclear reactors will be the central components in large electrical generating plants. In recent years, a growing portion of such generating capacity has been provided by nuclear power plants.

The nuclear power plants of this century depend on a particular type of nuclear reaction, fission, for the generation of heat. Fission is the splitting of a heavy nucleus, the center of an atom such as uranium, into two or more principal fragments, as well as lighter pieces, such as neutrons. In principle, fission may occur spontaneously, but in nuclear reactors this splitting is induced by the interaction of a neutron with a fissionable nucleus. Neutrons are, in fact, one of the two basic components of nuclei (the other is the proton) and, as noted, they are released during fission, thereby becoming available to induce subsequent fission events. Under suitable conditions, a "chain" reaction of fission events may be sustained. The energy released from the fission reactions provides the heat, part of which is ultimately converted into electricity. In present-day nuclear power plants, this heat is removed from the nuclear fuel by water that is pumped past the rods containing the fuel. Other fluids may be used as a coolant, but in every case this coolant delivers the heat, either directly or indirectly, to the electrical generating part of the plant. This mode of heat transfer does not differ in principle from that used in

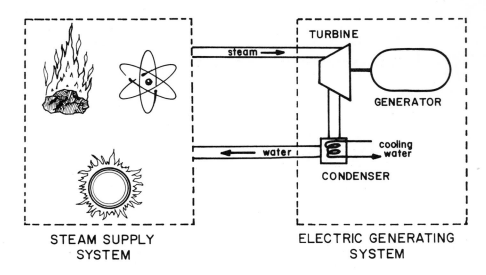

Figure 1-1. SCHEMATIC STEAM POWER PLANT.
In an ordinary electrical generating plant, heat from fossil-fuel combustion, from nuclear reactions, or even from the sun, is transferred to a supply of water under conditions suitable for boiling. The resulting steam drives a turbogenerator, producing electricity. Degraded steam is then condensed (on cooling by heat transfer to an external water supply) and returned to the boiler-reactor system.

plants that depend on chemical reactions, including the burning of coal, oil, or gas.

However, dependence on a fission chain reaction does introduce some special aspects to the reactor. The first arises from the fact that a nuclear reactor depends on a chain reaction. Maintaining a constant power level requires that the chain reaction be controlled so that, on the average, each fission causes only one subsequent fission. The second peculiarity of a nuclear reactor is that the products of reactor operation are highly radioactive. As a result, a substantial portion of the effort devoted to reactor design is aimed at limiting the probability of release of these products.

THE NUCLEAR CHAIN REACTION

The basic feature of a nuclear reactor is the release of a large amount of energy from each fission that occurs in the reactor's core. On the average, a fission event releases about 200 million electron volts[1] (MeV) of energy. A typical chemical reaction, on the other hand, releases on the order of one electron volt (eV). This difference, roughly a factor of 100 million, accounts for the fact that the complete fission of one pound of uranium would release roughly the same amount of energy as the combustion of 6000 barrels of oil or 1000 tons of high-quality coal.

Fissions are induced principally by the absorption of a neutron by a "fission-

[1] See "Miscellaneous Units and Equivalences," Appendix A.

able" nucleus. If, further, the nucleus can be induced to fission by very low-energy neutrons, the nucleus is referred to as "fissile" (see Glossary). Neutron absorption by a fissile nucleus results, a large percentage of the time, in the breakup of the nucleus into two major fission fragments, which are actually nuclei themselves, such as strontium or cesium. In addition, the fission produces less massive "pieces," principally neutrons and gamma rays (a high-energy form of electromagnetic radiation).[2]

On the average, over 80% of the energy released by fission is carried off by the fission fragments in the form of kinetic energy. These fragments are rapidly stopped by core materials and, in the process, their kinetic energy is converted to thermal energy, heating up these materials. Additional thermal energy results from stopping or absorbing the neutrons and gamma rays given off during fission, or the beta particles[3] and gamma rays which are subsequently released by the fission products (fragments).

The number of fissions is not only a convenient measure of the energy release in a reactor, but also a fundamental measure of the basic process in a nuclear reactor: fissions form the backbone of a nuclear "chain reaction." This chain is linked by fission-produced neutrons, which themselves produce fissions.

All nuclear reactors depend predominantly on fissile material for establishment of a chain reaction. The only naturally occurring fissile material is an isotope of uranium which has mass 235.[4] Uranium 235 constitutes only 0.7% of natural uranium; 99.3% is uranium 238. As noted in Chapters 2 and 11 and Appendix F, many reactors require that the uranium fuel be "enriched" to contain a higher percentage of ^{235}U. In commercial power plants, the fissile material is the source of almost all the fissions occurring in the reactor. When a nuclear power system is just being established, this fissile material must be ^{235}U. During reactor operation, other fissile nuclides, particularly plutonium 239 or uranium 233, are formed.

The character of a just self-sustaining chain reaction is suggested in Figure 1-2. As suggested by the figure, once a reactor is up to power, it is only necessary that, on the average, one of the neutrons produced per fission cause another fission. The number of neutrons in succeeding "generations" will then remain constant. This condition is called "criticality." Since an average fission event produces two or three neutrons, surplus neutrons are available and may be "captured" in a way that does not induce fission, either by the fuel itself or by the reactor structure and coolant. Alternatively, some neutrons are lost by leakage from the core.

---

[2] Gamma rays are only one form of electromagnetic radiation; light is another. Beta (β) decay amounts to the change, in a nucleus, of a proton to a neutron or vice versa, accompanied ordinarily by emission of a beta particle, a positively or negatively charged electron. Electrons have about 1/1836th the mass of neutrons or protons, the particles of which nuclei are constituted (see Footnote 4).

[3] See note 2 above.

[4] Uranium (chemical symbol U) and plutonium (Pu) are the elements that have, respectively, 92 and 94 protons in their nuclei. (The number of protons determines the element.) The particular "isotope" of an element is determined by the mass number, the total number of protons and neutrons. Uranium 235, the naturally occurring fissile isotope of uranium, is written in scientific notation as ^{235}U although, for convenience in text, it is often written U-235. Chemical symbols are given in Appendix A.

Figure 1-2. A SCHEMATIC NEUTRON ECONOMY.
Fission-produced neutrons can be absorbed in a number of different ways, of which ^{235}U fission is the one that continues the chain reaction. Between the various reactions shown in the figure, neutrons may be slowed down by collisions with moderating material. The economy is balanced if, on the average, one neutron is produced for each one absorbed. (Although each of the reactions shown has heavier products of some sort, we have for simplicity indicated only the resulting neutrons and gamma rays.)

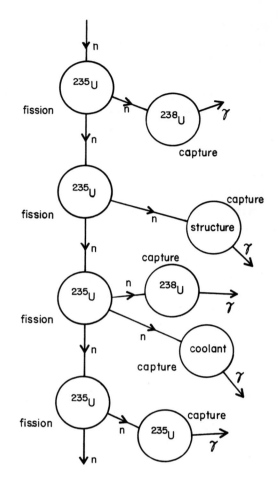

The probability that one or another type of interaction (between, say, a neutron and a nucleus) will occur is usually measured in terms of the "cross-section" for that particular interaction. (This concept is discussed at some length in Appendix B.) This probability (and the associated cross-section) depends on the particle types involved and on their state of motion with respect to one another. To cite an example, a slow neutron passing through a region containing fissile nuclei has a higher probability of inducing fission if the target nuclei are plutonium 239 than if they are uranium 235. This is consistent with the fact that, as noted in Table C-3,[5] the slow neutron cross-section for ^{239}Pu is larger than that for ^{235}U. The converse is true at certain higher neutron energies, where the neutron cross-section of both these nuclei is reduced, the ^{239}Pu by a larger amount.

The fact that at low neutron energy the fission cross-section is relatively large accounts for a basic design feature of currently available power reactors in this country: the fuel is surrounded by a "moderating" material which slows down

[5] Appendix C presents some basic characteristics of nuclear materials.

neutrons rapidly to improve their chances of inducing fission and thus continuing the chain reaction. The neutron energy is moderated by collisions with the relatively light nuclei of the moderating material, as discussed in Appendix B. The vast majority of nuclear power plants in the United States are based on "light-water" reactors (LWRs), in which the moderator is ordinary water; this water also serves as the coolant (see next section). Other choices of moderator are possible and, in some respects, preferable. To illustrate this point, we note that water (H_2O) contains hydrogen, which can "capture" a neutron, forming deuterium (and, incidentally, giving off a gamma ray γ, which has only a small amount of energy compared with that given off by a fission event). As a result, neutrons may be absorbed by the water with little energy yield. This loss of neutrons is not as severe if carbon or "heavy" water (in which the ordinary hydrogen is replaced by deuterium) is chosen as the moderator (see Appendix B). On the other hand, light water is both plentiful and dual purpose (it cools the core).

Neutrons are also captured (i.e. without fission) by the fuel material itself. This is not as unfortunate as it might seem; it can be turned to an advantage, actually producing fissile material. Capture of a neutron by uranium 238 or thorium 232 leads to the following series of reactions:

$$^{238}U + n \rightarrow {}^{239}U + \gamma \qquad {}^{232}Th + n \rightarrow {}^{233}Th + \gamma$$
$$^{239}U \rightarrow {}^{239}Np + \beta \qquad {}^{233}Th \rightarrow {}^{233}Pa + \beta$$
$$^{239}Np \rightarrow {}^{239}Pu + \beta \qquad {}^{233}Pa \rightarrow {}^{233}U + \beta$$

(See Appendix C for more details.) ^{239}Pu is itself fissile. ^{233}U is a third fissile nuclide of interest to nuclear power. Further, these nuclides are produced from ^{238}U and ^{232}Th by the capture of one neutron; neptunium (Np) and protactinium (Pa) are temporarily formed in these processes. In principle it is possible to build a reactor that produces more fissile material than it destroys. This is clearly advantageous, since it is the fissile material that yields most fissions and hence energy.

Thus there are two productive ends to a neutron: a fission event (yielding subsequent neutrons and a large amount of energy) and a conversion event (turning a "fertile" nucleus such as ^{238}U into a fissile nucleus, one that can be induced to fission by slow neutrons). The probability of such alternatives depends on the nuclide being considered. Various materials may be characterized by a number of properties (this list is far from exhaustive):

- The fission cross-section (both at thermal[6] and higher energies)
- The number of neutrons resulting from fission events (this will vary from one event to another, so that the number, like a cross-section, is actually an average)
- The capture to fission ratio (the ratio of the "capture" cross-section, wherein a neutron is absorbed and a gamma ray is given off, to the fission cross-section).

Appendix C contains additional information on nuclear materials.

[6] "Thermal" energy is the amount of energy that neutrons have after they have been permitted to collide a large number of times with their surroundings, which are at some particular temperature. The molecules constituting the surroundings will also have this same average energy (see Appendix B).

Figure 1-3. REACTOR CORE. An ordinary reactor core consists of long, thin, metal-clad fuel rods immersed in a cooling or moderating fluid. Each of the fuel rods, also called pins, is a stack of small cylindrical "pellets" of fuel material, currently uranium dioxide. The stack is surrounded by a metal can or "cladding," forming a rod, the basic fuel element in many reactors. Such fuel rods are typically assembled into bundles (not indicated in the figure), which are supported in a large reactor vessel through which the coolant is pumped.

Other ends to neutrons than fission or conversion (capture by a fertile nucleus) may be regarded as a loss. These include capture by moderator or coolant, reactor structural material, ^{235}U, and the products of fission (such as xenon 135). A careful reactor design minimizes these losses to the extent compatible with other requirements. Among the more notable of such requirements is the need to provide enough moderator to slow the neutrons down; this need must be balanced against the fact that an increase in the amount of water or other moderator means greater losses due to capture. Similarly, enough coolant must be provided to keep the core at safe operating temperature.

The design of a reactor requires careful treatment of two basic interacting quantities to which we have alluded: the neutron and heat fluxes, which are embedded in a common medium, the reactor core (see Figure 1-3). In general, a flux describes the flow of some material or other measurable, such as thermal energy. The quantities of direct interest in a reactor are the neutrons that permeate the core and the heat that is produced, primarily as a result of fission induced by these neutrons. It is clear that there is a strong connection between these quantities. Not only does the reaction rate affect the amount of heat that is generated, but the temperature, whose value is directly controlled by this heat and the manner in which it flows, affects the reaction rate. (See, for example, the "Doppler effect" discussed in Appendix E.)

Implicit in our discussion of the chain reaction is the fact that a neutron "born" at one point in space and time will eventually be absorbed at another point. The relationships or equations that describe the manner in which neutrons move from their point of appearance to their point of absorption contain information on all the ways in which neutrons can be created and destroyed, and on the interactions they can have in between, such as the collisions that "moderate" their energy. Moreover, for many purposes it is not sufficient for these equations to describe the materials that make up the core (these are the materials with which the neutrons interact) in only an average way. The equations must actually describe the spatial distribution of these materials and of the neutrons and thermal energy that exist in the materials. It is particularly clear that this spatial description must be employed at the boundary of a reactor core, since any neutrons that pass beyond this boundary, and are not returned by some sort of reflector, are lost to the system. In rough terms, this loss depends on the surface area of the reactor core, whereas the total number of neutrons being produced depends on the volume of the core. Therefore, because the ratio of surface area to volume decreases as the scale of the core increases, the relative importance of boundary losses decreases with core size. Because of this, for any given average core composition there will be a minimum size reactor for which criticality can be achieved, assuming that the composition is such that criticality is possible at all.[7]

Spatial variations are important not only at the boundary, but also within the core, since the core does not typically consist of a homogeneous mixture of fuel material, structural components, and coolant. The fuel is usually contained in rods

[7] The same consideration applies to nuclear weapons. See Chapter 12.

(long, thin cans) or the equivalent. Even the coolant is not uniform; water, for example, develops bubbles as it passes through the core, even in reactors that do not purposely allow the coolant to boil. Because of the heterogeneity, some fuel elements or portions of fuel elements may run hotter both in reaction rate and temperature than others. The spatial distribution of fissions or, alternatively, heat production is an important design consideration, affecting not only fuel burnup (the amount of energy that the fuel can release), but also the ability of the cooling system to cool the core adequately.

CORE COOLING AND ENERGY CONVERSION

The reactor cooling fluid serves a dual purpose. Its most urgent function is to remove from the core the heat that results when the energy released in nuclear reactions is transformed by collisions into random molecular motion (heat). An associated function is to transfer this heat for use outside the core, typically for production of electricity. The basic energy production, transfer, and conversion elements of light-water reactor nuclear generating systems are indicated in Figure 1-4. The designer provides for a nuclear core in a container through which a cooling fluid is pumped. This fluid may be used directly to drive a turbogenerator. Alternatively, it may be used to heat a secondary (or ultimately a tertiary) fluid which drives the turbine. In all present commercial systems, the final fluid is vaporized water, i.e., steam, although this is not the only possibility.

Figure 1-4 shows the pressurized-water reactor (PWR), in which the reactor cooling water is highly pressurized, so that it remains below the boiling point; the water leaves the reactor to pass through a "steam generator," where a secondary coolant (again water) is allowed to boil, producing steam to drive the turbine. Also shown is the boiling-water reactor (BWR), in which the steam produced in the reactor exits the reactor vessel and proceeds directly to the turbogenerator. More information on these reactors is given in Chapters 2 through 6.

The process of using the heat energy contained in the steam to provide the mechanical work to drive the turbine results in the rejection of most of the heat at a lower temperature to a waste heat sink, typically a body of water or a cooling tower. This heat transfer occurs at the condenser, where the steam from the turbine is converted to a liquid before it returns to the reactor vessel. As discussed in Appendix D, a low-temperature reservoir is necessary for any process that converts heat to mechanical work. The ratio of useful work (i.e. output electrical energy) to thermal energy from the reactor is the thermal efficiency of the reactor system (see Chapter 2).

The effectiveness of the coolant in preventing core overheating is directly related to the particular core design. In almost every reactor design in use or under consideration, the fuel material is in solid form, normally contained in metal cans along which coolant flows. (If not in cans, the fuel is maintained in some other matrix which may also serve a function other than structural, such as moderation. The only exception of note is the molten salt breeder reactor, which has a liquid core.) Most of the energy produced in the reactor is deposited in the fuel material

Figure 1-4. LIGHT-WATER
REACTOR POWER PLANTS.
Plants using ordinary water as the
reactor coolant may operate in two
distinct fashions: steam generated in
the reactor coolant may be used
directly to drive the turbogenerator,
or the coolant may recirculate
raising steam in a secondary system.
The second approach is that used in
"pressurized-water reactors" and in
reactors using other than a light-water
coolant. The first, and more direct,
approach is used in "boiling-water
reactors." In pressurized-water reactors, gross boiling in the primary
coolant is prevented by maintenance
of the system pressure at roughly
2250 pounds per square inch, about
twice the pressure in a boiling-water
reactor. (Figure reproduced from
ERDA-76-107.)

**PRESSURIZED-WATER
REACTOR**

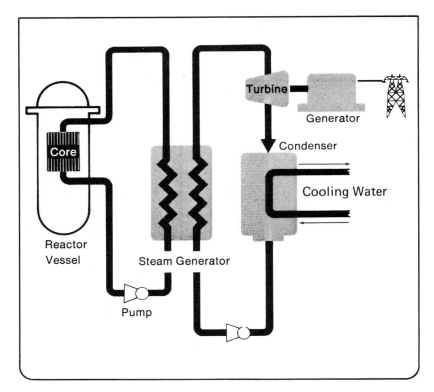

**BOILING-WATER
REACTOR**

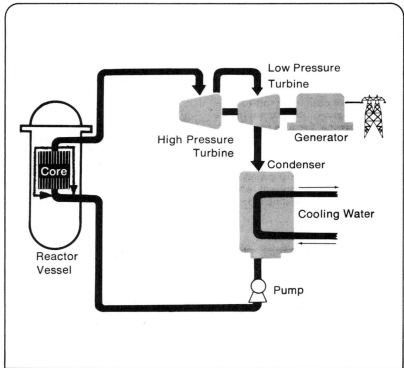

itself by fission fragments, since these fragments lose their energy in passing through a very small amount of material. The neutrons may deposit much of their energy in the moderator (in the case of thermal reactors), but they have relatively little energy to begin with (a few MeV as opposed to roughly 85 MeV per fission fragment); a similar comment applies to gamma rays. As a result, most of the heat from the reactor must flow from the fuel pellet to the surrounding material. Portions of the fuel will be substantially hotter than the nearby cooling fluid, and one of the prime design considerations is that the heat transfer be adequate to prevent temperature increases that may damage the fuel or its container. The parameters at the disposal of the designer are many, but the ones most obviously connected with this heat transfer requirement are the thermal conductivity of the fuel and surrounding material (the container, the coolant and/or moderator) and the character of the boundary layer between the coolant and the structure it cools. This boundary layer depends on flow rate, pressure, surface roughness, etc. Complicated as it is, it is the most basic and perhaps most variable element in the cooling system.

The complexities of nuclear reactor design contrast with the other end of the nuclear power generating system, the turbogenerator, where the use of steam to generate electricity is common to much of power engineering. In fact it has become conventional to divide the nuclear system into two parts (although this division may tend to obscure the important interdependences between the two parts): the nuclear steam supply system (NSSS), which includes the reactor itself and any associated equipment (such as steam generators) necessary to produce steam, and the turbogenerator system that is driven by this steam. Design of the NSSS is a task for the nuclear engineer; details of the heat-to-electricity conversion are handled on the basis of standard engineering thermodynamics and generator design. It is true, however, that basic changes in turbogenerator design may be prompted by requirements of a particular nuclear system. For example, it may be possible to achieve unusually high thermal efficiencies using a high-temperature gas-cooled reactor in conjunction with a gas (rather than steam) turbine. Design of a gas turbine that can withstand very high temperatures is at present being vigorously pursued. However, the vigor of these studies is also prompted by the potential for application of gas turbines to fossil-fueled systems.

REACTIVITY CONTROL

Keeping a reactor operating at a constant power level requires the maintenance of a delicate balance between neutron production and absorption. Roughly speaking, the rate of energy production, i.e., the power level, of a given reactor is proportional to the number of neutrons available in the reactor, and it is, therefore, essential that this number remain constant if the power level is to be stable. As a result, we require that, on the average, for every neutron absorbed or lost, precisely one be produced; the number of neutrons in succeeding generations will then remain constant.

It is straightforward to see how this balance can be *approximately* maintained. Neutrons are produced predominantly by fission events. (The few that arise

Figure 1-5. CONTROL RODS. The reaction rate can be controlled by "control rods," consisting of neutron absorbing material. Moving the rods alters the number of neutrons available for fission and hence controls the core reactivity. A reactor often has a special set of rods for fast shutdown of the chain reaction.

otherwise imply only a small correction to what follows.) Such an event, on the average, yields somewhat more than two neutrons. To maintain a constant power level, exactly one of these, on the average, must induce another fission event. The remainder must be absorbed in other ways. As we have indicated previously, the situation is happiest if most of these are invested in conversion of fertile material to fissile. Neutrons are the currency of the reactor economy, and as few of them should be wasted as possible. Some are lost through other processes, the most useful of which is absorption by control elements in the form of control rods or other mechanisms.

Why is control necessary? First, because it is not possible to design a reactor so that the number of neutrons in successive generations is *exactly* constant. Therefore, extra fissile material is included in the fuel, and control elements rob the system of enough neutrons to maintain a balance. This extra fissile material is needed, in any case, because over a long period of time, enough fissile material would be destroyed to turn the reactor off. Control mechanisms are needed because, in their absence, the extra fissile material at the beginning of operation would cause too great a portion of the neutrons to induce fission. As fissile material is "burned," control is deliberately withdrawn. This control is necessary for other reasons, the most prominent of which is the buildup of neutron poisons as a result of reactor operation — many of the nuclear reactions occurring in the core, in particular fission, produce nuclei that absorb neutrons with no useful result. Enough control must have existed when the reactor was turned on so that it can be reduced as this buildup of absorptive capacity occurs. A particularly important and interesting example of the need for control because of fission product poisoning is the so-called "xenon transient" to which reactors are susceptible during changes in power level (see Appendix E).

Having established the need for control, we can ask whether such control can be implemented. It is clear that introduction and removal of neutron absorbers can accommodate long-term needs for control (see Figure 1-5). However, the very short time between fission events in a chain reaction raises the question whether control elements can provide short-term stability to the fission rate. We can illustrate this question in terms of the "multiplication factor," the ratio of the number of neutrons in succeeding generations. In a thermal reactor, the time needed for a neutron to slow down and induce a subsequent fission (releasing neutrons) may be about 10^{-4} sec (one ten-thousandth of a second). Were this the actual time between all the fissions, we would have to conclude that, even if the multiplication factor k were only very slightly greater than 1, the power level would rise very rapidly. This is easily demonstrated: for k = 1.001, in only 1000 generations (taking a total of only 0.1 sec), the number of neutrons, and hence the power level, would rise by a factor of 1.001 to the thousandth power, or 1.001^{1000}, which is 2.7. This would be intolerable. On the other hand, moving control rods in a length of time much smaller than 0.1 sec does not seem feasible. A solution is provided, it turns out, by the simple fact that not all neutrons generated in the reactor appear "promptly." Most neutrons appear within times like 10^{-17} sec of the time of fission, which is what we mean by prompt. However, a small portion of them, about 0.5% for uranium-fueled thermal reactors, result from the decay of fission products which have half-lives (the time it takes for half of a sample to be transformed as a result of radioactive decay) on the order of seconds. These half-lives are so much greater than the 10^{-4} sec mentioned above that this small number of "delayed" neutrons substantially reduces the rate at which the neutron population changes. As a result, control can be achieved. (See Appendix E for more detail.)

One universal form of control is to provide a number of rods, loaded with neutron-absorbing elements such as boron or cadmium, which can be moved in and out of the core to select the portion of neutrons absorbed. In addition, reactors may have "burnable poisons" as part of the core — nuclei that absorb neutrons and, having done so, are neutralized. Moreover, poisons are commonly added to the cooling fluid. A number of other control mechanisms are sometimes available, particularly for use in emergencies, a need that is discussed in the next section.

EMERGENCY FEATURES AND CONTAINMENT

For every reactor, "normal operation" implies, among other things, that enough cooling is supplied to maintain the structural integrity of the reactor system, particularly the core, and that radioactivity generated in the core does not escape into the general environment. Within these two constraints, there is much latitude for a variety of designs. The same is true of the backup systems which come into play should some sort of abnormality occur. Even so, some general comments can be made. Before doing so, it is appropriate to point out that reactor safety features may be divided into two categories: *intrinsic* safety features, those that are inherent in the physical nature of the reactor concept being considered, and *engineered* safety features, systems that are added to the basic reactor concept. (See Figure 1-6.) An example of intrinsic mechanisms is that, in a light-water reactor, overheating of the coolant caused by an abnormal rise in the reaction rate tends to reduce the water density and thereby shut down the chain reaction due to insufficient moderation of neutrons. Alternatively, the emergency shutdown-control-rod system may be regarded as an engineered safety feature. A specific reactor design will attempt to maximize the safety advantages offered by intrinsic physical characteristics and back them up by independent engineered systems.

In the discussion of cooling, we noted that the fuel pellet itself sustains the highest temperatures in the reactor. Under normal operation, this temperature is always well below the melting point of the fuel. For many reactors, however, it is the melting point of the fuel can or "cladding" that is more critical, since the cladding may melt at a lower temperature than the fuel and it is the cladding that provides structure to the core (especially in reactors where the core is cooled by a liquid). Should the melting point of the cladding be approached, fuel rods may bend or expand enough to disturb the coolant flow pattern, possibly accelerating, if only locally, the difficulty of cooling. Actual disintegration of the can would be even worse in this respect, and could also result in release of large amounts of

Figure 1-6. EMERGENCY
FEATURES FOR LIGHT-WATER
REACTORS.
The shutdown systems and emergency core cooling functions are
designed to prevent fuel melting. The
containment and its subsystems are
designed to limit releases of radioactivity, both from minor accidents,
and from those that involve major
core melting.

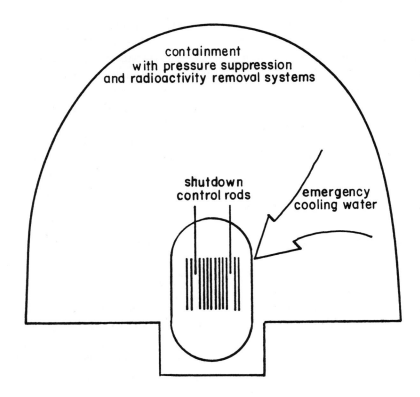

radioacitivity into the cooling fluid. For these reasons, the cladding should remain
functional at temperatures considerably higher than the design operating temperature.

The cladding provides, in addition to structure, one of the important barriers
against radioactive releases. The large quantity of radioactivity generated as the
reactor runs must be confined at the reactor site. The most obvious means is to
keep it in the fuel material itself. To a large extent, this occurs automatically since
most nuclear reactions occur in the fuel and the fuel is a solid. The reaction
products, particularly the large fission fragments, tend to lodge in the fuel material.
However, some of the reaction products are gaseous, especially at typical operating
temperatures, and some portion of these will permeate the fuel material. The cladding (or other fuel coating) can prevent escape of such radioactivity from the fuel
to the surroundings. To prevent release of gases, it is necessary for the metal
cladding (that is used in light-water reactors) to be welded shut, then leak checked.

In spite of welding and checking, a small percentage of the fuel pins or rods
leak gases into the reactor coolant. Moreover, some radioactivity is generated in the
coolant itself. (For example, a small portion of the tritium — the mass 3 isotope of
hydrogen — released from light-water reactors is formed by capture of neutrons by
minor amounts of deuterium in the water. For some systems, it is significant that
neutrons may produce ^{16}N from the oxygen in the water coolant). Regardless of its
source, long-lived radioactivity may be removed from the coolant by a cleaning
system of some sort. The collected material may then be saved for later disposal or

released in a controlled manner into the general environment (see Chapters 2 and 3).

Although a generic description of systems for dealing with abnormalities is much more difficult, a few examples would be useful. The most basic system is incorporation of "control" as discussed in the last section. In many reactors, there are both "control" and "shutdown" rods, the latter being specifically designed for rapid and failsafe shutdown of the chain reaction. The actual speed with which shutdown occurs (i.e., the rate at which the fission rate decreases) depends on the amount and rate of control added. A normal operational change in power level may occur over a period of hours. An emergency shutdown may take seconds, a significant time interval in some circumstances.

As mentioned before, there may also be recourse to dumping neutron absorber in other forms into the reactor. (In some cases, this technique is also used for routine control.) It is generally assumed that there are enough backup systems to guarantee shutdown when it is necessary.[8] Some abnormalities of themselves guarantee shutdown. The most notable of these is that, if the cooling water from a light-water reactor is lost, the reactor is reduced below criticality because neutrons are no longer moderated to the low energies where the fission cross-section is higher. As a result, the chain reaction comes to a halt, even without the insertion of control rods. For fast reactors, however, loss of the coolant does not preclude criticality; it may have the opposite effect, increasing the reaction rate. For any type of reactor, continued cooling is still required after shutdown.

We cannot pretend to outline all the circumstances that would warrant a shutdown (called a "scram" if conditions require a very rapid shutdown). Ordinarily, any abnormality (operation out of predetermined tolerance) in operational parameters, such as coolant temperatures or flows, fuel pin temperatures, neutron flux, or any abnormality in the instrumentation checking these quantities, or any malfunction of a system that would be needed in an emergency, may provoke shutdown of the reactor. As a result, most shutdowns are not due to malfunctions of the basic reactor system, but rather of support and monitoring systems.

When the chain reaction is stopped by insertion of control, it is still not feasible to shut down all other reactor functions. Most importantly, the coolant system must continue to run because the reaction products continue to generate heat through radioactive decay. Immediately after shutdown, the power level is still about 7% of the level that was sustained before shutdown, and this 7% is enough to require continued cooling of the core in order to prevent damage to the fuel from overheating. For this reason, some portion of the reactor cooling capacity must continue to be available, although this need only be a small portion of the capacity needed for full power operation.

It is, therefore, necessary that cooling be available under all conditions. In particular, even if the reason for the shutdown is loss of cooling, enough must

---

[8] This guarantee may not be unambiguous in some cases such as the fast breeder reactor, where the possibility of reassembly of damaged or melted fuel into a critical configuration must be considered. Reliable shutdown systems for breeders such as the liquid metal fast breeder reactor are in the process of development.

remain or be supplied to prevent serious damage to the core. In the absence of cooling, portions of the core would actually melt, thus resulting from the operational point of view in destruction of the reactor, and possibly leading to a serious release of radioactivity. Because of the consequences of failure, components in the cooling system may be duplicated and have independent sources of power. Moreover, in light-water reactors, there is an emergency core cooling (ECC) system that is largely independent of the main system. This philosophy of redundancy and independence is typical of any reactor design. It is particularly important in the shutdown and cooling systems. Even so, adequacy of these systems has been a matter of dispute.

To conclude this section, we should note that, to some extent, the possibility of breach of the reactor's primary system itself must be considered. This could come in the form of a break in a cooling line or even in the reactor vessel. Emergency systems are provided to cope with the more probable of such events. The first line of defense is, in fact, the ECC system. In addition, the reactor vessel and primary cooling system will typically be enclosed by a containment building designed to withstand a significant overpressure. Inside this building will be systems to prevent substantial overpressure and to clean the internal atmosphere of radioactivity before any possible release to the external environment. The most serious, but presumably least likely, event would be a core meltdown that breached the reactor vessel, then the containment building. Various predictions have been made of the course of such an event, but it is difficult to specify a precise sequence for such an accident; some protection would be afforded by the filtering action of the materials underlying the building. The potential for serious accidents is discussed in Chapter 4.

Many of the systems mentioned above are examples of engineered safety features, features designed to limit the extent of any abnormality and to mitigate radioactive releases and damage to the reactor system from operation under both normal and abnormal conditions. Consideration of the possibility and probability of various events is an intimate part of reactor design. It is fair to say that reactor designers are typically satisfied that they have sufficiently provided for the safety of the public and of the reactor itself. Whether others are so satisfied is another matter.

Bibliography — Chapter One

"NTIS" indicates report is available from: National Technical Information Service, U.S. Department of Commerce, 5285 Port Royal Road, Springfield, VA 22161.

APS-1975. H. W. Lewis, et al., "Report to the American Physical Society by the Study Group on Light-Water Reactor Safety," *Reviews of Modern Physics,* vol. 47, supplement no. 1, p. S1 (1975).
    Reviews the main features of light water reactors, their safety features, and the U.S. reactor safety research program.

CONF-740501. "Gas-Cooled Reactors: HTGR's and GCFBR's," topical conference, Gatlinburg, May 7, 1974 (NTIS).

A broad selection of technical papers on various aspects of gas-cooled reactors.

ERDA-76-107. "Advanced Nuclear Reactors," U.S. ERDA report ERDA-76-107, (May 1976) (NTIS).

A brief and elementary introduction to advanced reactor types.

Foster, J. S., and Critoph, E., "The Status of the Canadian Nuclear Power Program and Possible Future Strategies," *Annals of Nuclear Energy,* vol. 2, p. 689 (1975).

A brief description of possible configurations of the CANDU reactor.

Glasstone, S., and Sesonske, A., *Nuclear Reactor Engineering* (Van Nostrand Reinhold, New York, 1963).

A thorough introduction to nuclear engineering.

Lamarsh, J. R., *Nuclear Reactor Theory* (Addison-Wesley, Reading, Mass., 1966).

A quantitative treatment of the physics of nuclear reactors.

Penner, S. S., *Nuclear Energy and Energy Policies,* vol. 3 of *Energy* (Addison-Wesley, Reading, Mass., 1976).

A general introduction to the various kinds of nuclear-generated energy, including both fission and fusion systems.

Seaborg, G. T., and Bloom, J. L., "Fast Breeder Reactors," *Scientific American,* vol. 223, p. 13 (November 1970).

A description of fast breeder reactors, including the LMFBR and the GCFR.

Steam, Its Generation and Use, 28th ed. (Babcock & Wilcox, 1972).

A detailed treatment of many of the engineering aspects of steam systems, including both nuclear and fossil-fired generating systems.

WASH-1250. "The Safety of Nuclear Power Reactors (Light-Water Cooled) and Related Facilities," U.S. AEC report WASH-1250 (July 1973) (NTIS).

A broad and useful description of light-water reactors and of their safety features, from the point of view of the Atomic Energy Commission.

Wessman, G. L., and Moffette, T. R., "Safety Design Bases of the HTGR," *Nuclear Safety,* vol. 14, p. 618 (1973).

Describes HTGRs and their safety considerations.

Wilson, R., and Jones, W. J., *Energy, Ecology, and the Environment* (Academic, New York, 1974).

An elementary treatment of the environmental impact of energy-producing systems.

CHAPTER TWO

The General Environmental Interaction

NUCLEAR POWER PLANTS and associated facilities have a noticeable interaction with the rest of the world. This hardly needs saying. After all, these plants are constructed to provide an important utility, electricity. But it is a point worth emphasizing and clarifying. The last chapter discussed basic design features of reactors, but paid little attention to the "environmental" interaction. This chapter serves as an introduction to this interaction in its broad sense: the resource-utilizing and energetic, as well as the narrowly construed environmental, health, and safety aspects of nuclear systems.

INPUT-OUTPUT PARAMETERS

We have already alluded to two of the environmental emissions from nuclear power plants: radioactivity and heat. However, a more general view identifies a larger array of "substances" which pass into and out of reactor systems. Many of these are shown in Figure 2-1. The flows associated with an existing power plant may be characterized as fuels, energy, and emissions. Additional commitments, such as land and the materials for building the power plant, may also be important, but are not considered in this book. However, the energy required for building the power plant is considered briefly at the end of this chapter.

Figure 2-1 shows that the basic input to the reactor is fissionable material, in the form of fresh fuel, and the primary outputs are electricity and irradiated fuel. Only a portion of the heat generated in the reactor is transformed into electricity, the remainder being rejected to the environment through the condenser cooling system. Moreover, not all the radioactivity contained in the reactor is removed from the nuclear power plant in the form of irradiated fuel; a small portion is released to the general environment as liquid and gaseous wastes (see chapter 3), while still other materials are stored in solid form. This reject heat and radioactivity are the important direct environmental "residuals" from the operation of a nuclear power

18

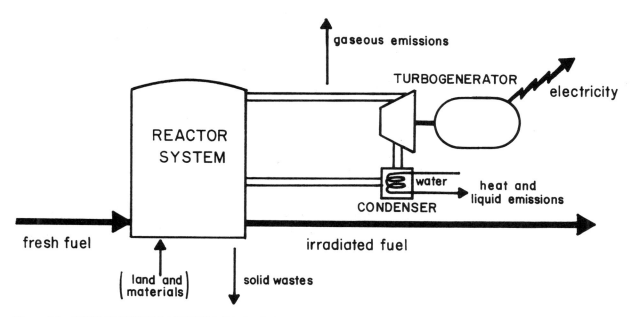

Figure 2-1. SCHEMATIC INPUT-OUTPUT CHARACTERISTICS OF A NUCLEAR POWER PLANT.
The primary input to a nuclear power plant is fresh fuel; the primary outputs are electricity and irradiated fuel. There are, in addition, other material and energy flows into and out of the plant. The most important of these are the radioactive species contained in the gaseous and liquid emissions.

plant. The irradiated fuel is regarded as a primary output because it inevitably contains fissionable materials which may be worthy of recycle.

It is, in fact, the fuel stream, including both fresh and irradiated fuel, that connects the nuclear power plant to the remainder of the "nuclear fuel cycle." The so-called "front" end of the fuel cycle includes all the facilities necessary for the production of fresh fuels. These are the resource extraction operations and the facilities that convert the raw uranium or thorium into the fuel form appropriate for a given reactor type. The "back" end of the fuel cycle includes all the facilities necessary for the handling, processing, or storage of irradiated fuel or of materials extracted from the fuel. Fuel fabrication facilities which utilize recycled uranium, plutonium, or thorium, should such recycle occur, would serve to connect these two portions of the fuel cycle. The chemical dissolution of the irradiated fuel and the separation of the resulting solution into fuel and waste fractions is referred to as "reprocessing." (This term may also be taken to include conversion of the fuel and waste fractions into some particular chemical and physical form.) The manner in which the nuclear power plant would be embedded in a simple fuel cycle is indicated in Figure 2-2. More complex cycles result when the possibility of exchange of fuel materials between two or more types of reactor systems is considered. In most of this book the fuel cycle is regarded only as a source and repository for fuel. A later chapter (11) is devoted specifically to the questions of fuel reprocessing and waste management. Appendix F summarizes the chemical and physical processes

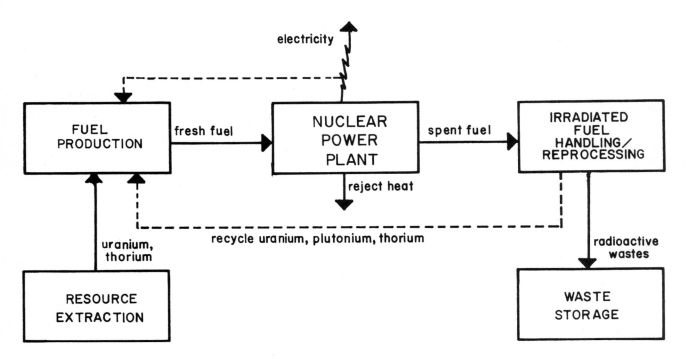

Figure 2-2. GENERALIZED NUCLEAR FUEL CYCLE.
The basic function of a nuclear power system is to extract uranium or thorium resources and produce electricity. The figure indicates the main flows of fuel resources and energy for the nuclear fuel cycle. In many nuclear systems, recycle of fuel materials is not necessary. A similar, although somewhat simpler, scheme can be associated with a fossil-fuel cycle.

usually associated with the light-water reactor fuel cycle; there the environmental questions associated with the fuel cycle are indicated. These are also considered briefly in the next chapter.

It is clear that some input-output aspects of nuclear power can be considered on a plant basis, while others need to be considered in the context of the entire fuel cycle. In many cases, for example, the environmental residuals — heat and radioactivity — would need to be considered in the context of a particular plant site in order to assess the impact and hence importance of these residuals. On the other hand, uranium utilization characteristics are inextricably connected with the fuel cycle design. So, although we will often associate uranium requirements with a particular reactor type, some sort of supporting fuel cycle will be implied and, it is hoped, obvious.

Finally, certain input-output parameters, whether of the nuclear power plant or of the entire fuel cycle, have their full implications only when considered in an even broader context. For example, the electricity produced by a nuclear power system can be valued only on the basis of the perceived need for electric power. As a further example, the importance of the radioactive emissions (whether routine or accidental) from a nuclear power plant or from fuel cycle facilities can be evaluated only on the basis of an understanding of the relationship between exposures to radiation and harmful effects to humans and of the negative value of these effects. As a final example, the implications for weapons proliferation of the spread of commercial nuclear power can be considered only in the broadest context.

The emphasis in this book is on the characteristics of the nuclear power plants themselves. The nuclear fuel cycle is considered in sufficient detail to establish the context in which the basic reactor features must be considered. Finally, some of the material in Parts I and III establishes a limited technical basis for making judgments on the role of nuclear power.

CHARACTERISTICS OF THE PREDOMINANT REACTOR TYPES

It is useful at this point to summarize briefly the main features of some of the reactor types that have been available commercially. The reactors considered here are the pressurized-water reactor, the boiling-water reactor, the heavy-water reactor, and the gas-cooled reactor. All the types considered are "thermal" reactors, that is, the neutrons are moderated to take advantage of the elevated fission cross-sections at low neutron energy. The physical design features are treated in more detail in Part II. The primary interest in the present discussion is the input-output characteristics of these reactors, but we will indicate their general concept as well.

In the United States, virtually all commercial nuclear power is generated by light-water reactors. Commercial power plants are shown in Figure 2-3, all but the one in Colorado being LWRs. These reactors use ordinary water, in which the hydrogen is almost entirely the mass 1 isotope, as the moderator and coolant. This hydrogen has a high enough probability of capturing a neutron to form deuterium, the mass 2 isotope of hydrogen, to represent a significant loss of neutrons to the system. The probability is so severe, in fact, that natural uranium cannot be used as the fuel in a light-water reactor. All the fuel is, indeed, uranium. But the concentration of the fissile isotope uranium 235 has to be increased from its natural 0.7% to almost 3% in order for such a reactor to operate. This "enrichment" is a difficult, expensive and energy-consuming process, the main reason, in fact, that Figure 2-2 indicates electrical power being returned in significant amounts to fuel production facilities.

The light-water reactor is found in two types, the pressurized-water reactor and the boiling-water reactor. The heat transfer systems for these reactors were displayed in Figure 1-4. The BWR is somewhat simpler conceptually than the PWR in that steam is actually produced in the reactor vessel itself, and this steam drives the turbogenerators. In the PWR, the coolant is kept under greater pressure, so that substantial boiling does not occur. Instead, the primary coolant is used to raise steam in a secondary coolant, which proceeds to the turbine. Most commercial nuclear power plants in this country use pressurized-water reactors, sold by Westinghouse, Babcock & Wilcox, and Combustion Engineering. Substantial numbers use boiling-water reactors, sold by General Electric.

Some basic parameters for 1000-megawatt (electrical) (MWe) versions of

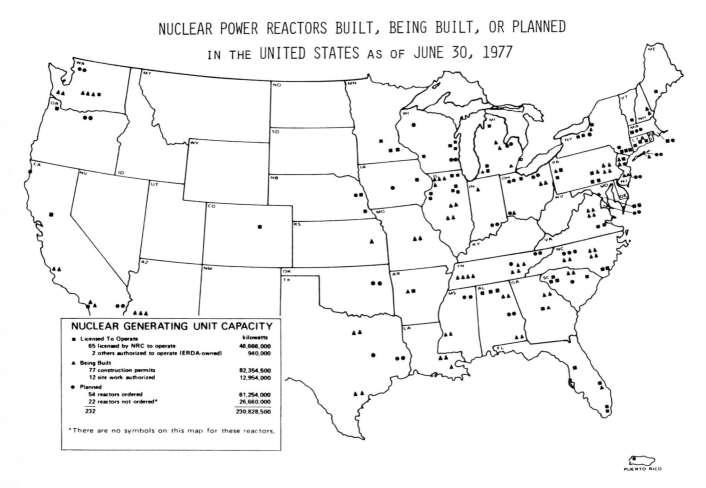

NUCLEAR POWER REACTORS BUILT, BEING BUILT, OR PLANNED
IN THE UNITED STATES AS OF JUNE 30, 1977

NUCLEAR GENERATING UNIT CAPACITY

	kilowatts
■ **Licensed To Operate**	
65 licensed by NRC to operate	46,666,000
2 others authorized to operate (ERDA-owned)	940,000
▲ **Being Built**	
77 construction permits	82,354,500
12 site work authorized	12,954,000
● **Planned**	
54 reactors ordered	61,254,000
22 reactors not ordered*	26,660,000
232	230,828,500

*There are no symbols on this map for these reactors.

Figure 2-3. NUCLEAR POWER REACTORS IN THE UNITED STATES.
The figure shows nuclear plants that are operating, under construction, or planned in the United States, as of June 30, 1977. (Figure reproduced from TID-8200-R36, "Nuclear Reactors Built, Being Built, or Planned in the United States as of June 30, 1977," available from National Technical Information Service.)

these reactors are shown in Table 2-1. These parameters include design features of the coolant and fuel systems, fuel utilization characteristics, and thermal parameters. For the light-water reactors, the amount of uranium that must be fed to the nuclear fuel cycle from sources of natural uranium is given, presuming that fuel reprocessing occurs and that reclaimed uranium and plutonium (which is produced primarily from neutrons captured by ^{238}U), is recycled to the fuel fabrication facilities. Reprocessing and recycle are not now occurring, but have until recently been considered to be part of the ultimate form of the LWR fuel cycle. As discussed in Part III, it is not necessary to reprocess or recycle for current reactors. For LWRs, the recycled uranium and plutonium each reduce the natural uranium feed requirements by about 15%. The fuel requirements given in Table 2-1 assume this reduction and give annual equilibrium commitments, as well as lifetime commitments for each reactor, including both the initial core and replacement loadings. (Comparable requirements for advanced reactors are given in Chapter 10.) These requirements assume that, when enrichment occurs, the uranium fraction that is depleted (rather than enriched) in uranium 235 contains only 0.2% ^{235}U, less than the percentage at which enrichment plants have been operated during the 1970s. A

TABLE 2-1.

Representative Characteristics of Commercial Power Reactors (1000-MWe capacity)

	PWR Pressurized-Water Reactor	BWR Boiling-Water Reactor	CANDU Canadian Deuterium-Uranium	HTGR High-Temperature Gas-Cooled Reactor
Coolant	Ordinary water (H_2O)	Ordinary water (H_2O)	Heavy water (D_2O)	Helium gas
Moderator	Ordinary water	Ordinary water	Heavy water[a]	Graphite
Percent ^{235}U enrichment	2-4%	2-4%	0.7%	93% (initial load)
Fertile (bred) nuclide	$^{238}U(^{239}Pu)$	$^{238}U(^{239}Pu)$	^{238}U (^{239}Pu)	^{232}Th (^{233}U)
Yearly uranium requirement at equilibrium (tons of U_3O_8)[b]	129	121	125	85
Lifetime uranium requirements (tons of U_3O_8)[b]	4,100	4,020	4,160	2,980
Thermal efficiency (percent)	32-33%	33-34%	28-30%	39%
Approximate once-through external cooling water requirements (gal/min with 15 °F temp. rise)	1,000,000	960,000	1,220,000	740,000
Core type	Fuel rods (bundled into assemblies)	Fuel rod assemblies	Fuel rod assemblies (individually pressurized)	Fuel particles dispersed in graphite blocks
Coolant pressure, psi (MPa)	2,250 (15.5)	1,020 (7.0)	1,490 (10.3)	700 (4.8)
Coolant temperature at exit from core, °F (°C)	620 (327)	545 (285)	590 (310)	1,370 (743)

a. The D_2O moderator is separate from the coolant and is at essentially atmospheric pressure (15 psi).
b. These uranium requirements are abstracted from Table 10-1. They assume recycle of plutonium and uranium for the LWR and BWR, recycle of uranium for the HTGR, and no recycle for the CANDU. (The requirements of present CANDUs are higher than given here.)

higher percentage in these "tails" implies a larger amount of natural uranium feed in order to yield the required amount of slightly enriched (i.e., 2% to 4%) uranium. In the discussion in later chapters, it is useful to remember that the lifetime uranium requirement of a light-water reactor is 4000 to 6000 tons of natural U_3O_8, the chemical form in which uranium is sent from the mining and milling operation. About 10% of this total has to be supplied for production of the initial core. For more details, see Part III (Tables 10-1 and 10-2).

Parameters for two other types of reactor are given in Table 2-1. The first is an example of a reactor that uses "heavy" water as the moderator and coolant. As noted earlier, the deuterium in heavy water captures substantially fewer neutrons than the hydrogen of "light" water. This smaller loss of neutrons means that a smaller percentage of fissile material is needed to reach criticality. (There is less neutron absorbing material to compete for neutrons with the fissile nuclei.) As a result, the fissile content of natural uranium is sufficiently high to achieve criticality with this type of moderator. Such a reactor is sold by Atomic Energy of Canada Limited, both in Canada and abroad (although not in the United States). The reactor that they market at present, called the CANDU (for "Canadian deuterium-uranium" reactor), actually has separate moderator and coolant systems. The moderator fills a large low-pressure vessel, through which are built a large number of individually pressurized and cooled fuel channels. As a result, the reactor has a noticeably different structure from an LWR. Heated heavy water from the pressurized channels runs through steam generators, producing ordinary steam for driving the turbogenerators.

The CANDU needs no fuel enrichment facilities. Moreover, the fissile content of the irradiated fuel is so low that it has ordinarily been presumed that uranium and plutonium would not be reclaimed from the fuel. The lifetime uranium feed for this reactor is still comparable to that of light-water reactors utilizing recycled uranium and plutonium. This is true in spite of the fact that the thermal efficiency (the ratio of electrical energy output to thermal energy generated) of the CANDU is lower (28%-30%) than that of LWRs (32%-34%).

We also show parameters for the high-temperature gas-cooled reactor (HTGR) designed, but not now marketed, by General Atomic. The water-cooled reactors discussed above have fuel pins bundled into assemblies which are then immersed in water. The HTGR is altogether different, having a core consisting primarily of stacked blocks of carbon. Small regions of each block contain fuel particles,[1] individually coated with hard ceramic. The carbon moderates neutrons, and the core is cooled by helium gas which flows through and around the carbon blocks. The heated helium raises steam, which powers the turbines, as in water-cooled designs. The core, the steam generators, and the coolant pumps are all contained in a single concrete vessel, causing this reactor to have a very different layout from the others described.

The fuel used in this reactor is radically different from that used in the water-cooled reactors. The fuel design uses thorium 232 (the naturally occurring

---

[1] This structure of carbon blocks and clumps of fuel is reminiscent of Fermi's 200-watt reactor of 1942.

isotope) as the fertile material, rather than uranium 238. The uranium fed into the reactor is highly enriched in uranium 235 (93%), thereby eliminating most of the uranium 238. The ^{233}U produced by neutron capture in the thorium fuel particles is recycled to the reactor after fuel reprocessing. (Any plutonium produced from the uranium feed would be relegated to waste storage.) Note that the lifetime uranium requirements for the HTGR are less than that for the other reactors. However, note also that this is due in substantial part to the higher thermal efficiency of the HTGR: for any specified amount of heat produced per ton of uranium feed, the HTGR would yield a greater electrical output.

These four reactors all depend on ^{235}U for their initial fissile loading. As indicated in Table 1-1, the "enrichment," the percentage of ^{235}U, varies considerably. However, in each case the uranium is present in a ceramic form, often as uranium dioxide (UO_2), but also — in the case of the HTGR — as a carbide. Use of uranium dioxide, rather than metallic uranium, reduces the density of uranium atoms in the fuel and also reduces the heat conductivity of the fuel, both disadvantages from the point of view of reactors. However, ceramics such as uranium dioxide have the advantage over metallic uranium of greater stability against radiation damage and chemical corrosion; the dioxide also retains a large portion of the gaseous fission products which in other environments might more easily escape. Uranium dioxide's high melting point (2800 °C) allows high-temperature operation, overcoming to some extent its low heat conductivity. The other nuclear materials with possible importance in reactors are plutonium and thorium. For our "commercial" nuclear reactors, these too would be present as dioxides (and carbides). However, it is important to note that advanced forms of these reactors could utilize nuclear fuels in other chemical forms.

We do not at this point attempt to characterize in any detail the radioactive emissions, either routine or accidental, of nuclear power plants. These emissions are the subject of the next two chapters.

OPERATIONAL FLEXIBILITY

A major difficulty in attempting to characterize reactors is that their input-output parameters can vary substantially, depending on how they are built and operated. Perhaps the most fundamental of such parameters is the amount of uranium required to produce a specified output of electrical energy. This required feed will clearly depend on whether irradiated fuel is being reprocessed to recover the remaining fissile materials. However, for currently available commercial reactors, all of which are water cooled, this dependence is not large: the natural uranium feed per electrical kilowatt-hour (kwh) is the same to within a factor of 2. This is true because the energy liberated from a fission event is largely independent of the type of nucleus fissioning; the ratio of fissile material produced to that consumed (the "conversion ratio") does not vary greatly among these reactors at the present time; and the efficiency with which heat is converted to electricity is always about one-third for present reactors.

For light-water reactors, in particular, it is possible to cite a characteristic set

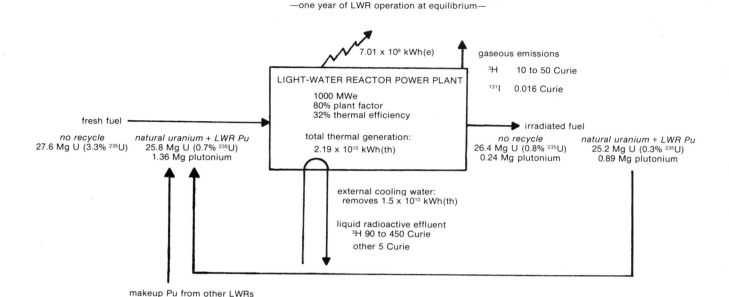

MATERIAL FLOW AND ENVIRONMENTAL RELEASE
—one year of LWR operation at equilibrium—

7.01 x 10⁹ kWh(e)

gaseous emissions

³H   10 to 50 Curie

¹³¹I   0.016 Curie

LIGHT-WATER REACTOR POWER PLANT

1000 MWe
80% plant factor
32% thermal efficiency

total thermal generation:
2.19 x 10¹⁰ kWh(th)

fresh fuel

irradiated fuel

no recycle
27.6 Mg U (3.3% ²³⁵U)

natural uranium + LWR Pu
25.8 Mg U (0.7% ²³⁵U)
1.36 Mg plutonium

no recycle
26.4 Mg U (0.8% ²³⁵U)
0.24 Mg plutonium

natural uranium + LWR Pu
25.2 Mg U (0.3% ²³⁵U)
0.89 Mg plutonium

external cooling water:
   removes 1.5 x 10¹⁰ kWh(th)

liquid radioactive effluent
   ³H 90 to 450 Curie
   other 5 Curie

makeup Pu from other LWRs

Figure 2-4. MATERIAL FLOW AND ENVIRONMENTAL RELEASE FOR LIGHT-WATER REACTORS.
The quantities shown are approximate fuel and emission quantities for a 1000 MWe LWR operating, on the average, at 80% of rated capacity. It is assumed that the reactor has operated long enough for the fuel inputs and outputs to have reached equilibrium. The "no recycle" data could apply equally well to the case of uranium recycle, where recycled uranium would decrease the amount required from mineral resources. The "natural uranium + LWR plutonium" case assumes the feed to the reactor is natural uranium supplemented by plutonium from uranium-fueled reactors. See Table 11-2 and Figures F-1 and F-2 for more detailed information.

of input-output parameters, with some allowance for the possibility of recycle of uranium and plutonium. These are given in Figure 2-4. The flow of fuel materials is that which would occur after the reactor had operated for some time. Because the initial core differs from the average lifetime core composition, the flows into and out of the reactor will differ noticeably from the equilibrium values for an initial period. This type of figure is useful because it fits conveniently into a comparable flowsheet for the entire fuel cycle (see Appendix F). The figure, in fact, not only gives the material flow and environmental releases for the power plant itself, but indicates the overall fuel cycle flow of nuclear materials.

Of the three reasons given above for similarity in fuel requirements, only the second has a great potential for change. Major changes in the type of reactor or in the associated fuel cycle cannot affect the relative invariance of the energy release per fission. Nor is there hope for a radical improvement in energy conversion efficiency. There is, however, substantial room for improvement in the "conversion ratio," the ratio of the amount of fissile material produced (primarily from neutron capture by uranium 238 or thorium 232) to fissile material destroyed (primarily from fission). The fissile material produced in the reactor can contribute directly to the production of power in the reactor in which it is produced; alternatively, it can be extracted from the irradiated fuel for later use in the same or other reactors. If as much fissile material is produced as is destroyed (i.e., the conversion ratio is 1), the reactor could run indefinitely on one fuel load, except for the loss of fertile material (due to conversion), the buildup of neutron absorbing fission products, and other effects. (With fuel reprocessing and the availability of fertile material, these difficulties can be side-stepped.)

Reactors that are designed to achieve high conversion ratios or even to produce more fissile material than they consume are treated explicitly in Parts III and IV. However, even for currently available reactors, the conversion ratio may be strongly influenced by simple design and operational factors, such as the frequency with which a portion of the core is replaced with fresh fuel. Replacing one-sixth of the core every six months, instead of one-third yearly, would reduce the amount of control needed; with less control, fewer neutrons are absorbed without benefit, and the conversion ratio rises.

Altering other reactor parameters can improve uranium utilization. Any improvement in the thermal efficiency of the power plant, the percentage of heat energy that is converted to electrical energy, reduces the required amount of uranium feed. This efficiency is often fixed by materials limitations, i.e., their ability to withstand the higher temperatures that would give better efficiency (see Appendix D). However, aside from these basic constraints, a less than optimum efficiency may be dictated by external constraints, such as a requirement that the low temperature heat sink be a dry cooling tower.

Choosing how any reactor is actually operated depends on a complex balance of cost, technical feasibility, and environmental requirements. A reactor vendor may, for example, design the fuel (and thus specify the conversion ratio) to obtain a balance of short-term and long-term costs that best suits the financial situation of the utility purchasing the reactor. Raising the conversion ratio, thereby decreasing the amount of uranium required for the lifetime of the reactor, may require increasing the amount of uranium (and hence money) initially invested in the reactor. The lifetime uranium requirements given in Table 2-1 are typical of reactor specifications as they have been. For a given reactor type, variations around these numbers have not been large, except that greater uranium feed has been required because reprocessing has generally not been available. In later chapters, we will discuss these particular reactor types in more detail, pointing out how alterations can affect their fuel utilization and other characteristics.

THE ENERGY INVESTMENT IN A REACTOR SYSTEM

Since the purpose of a nuclear power system is to make energy stored in the nucleus available as electricity, it is important that more energy be produced by the system than is invested in the construction or operation of the system. Two aspects of commercial nuclear reactors have led in the past to sometimes acrimonious questioning of whether the net energy from nuclear power operations is positive.

The first is that a fission power plant is an example of relatively high technology (not as high, however, as a fusion power plant would be). Actual construction of the plant and, to a lesser extent, of support facilities requires a substantial investment of labor and materials, both of which have associated energy expenditures. The second peculiarity of commercial power plants in this country is that they require that the fuel be enriched in its fissile content above the natural uranium value of 0.7% ^{235}U. The enrichment process used at present is a "gaseous diffusion" process in which uranium hexafluoride is heated to a gaseous state and pumped through porous barriers through which the mass 235 isotope passes more readily. The "separative work" required to produce fuel for a light-water reactor requires about 4% or 5% of the output of the power plant on the average, a substantial amount.

Both the construction energy investment and a substantial portion of the separative work occur before the operation of the nuclear power plant. How rapid a growth of nuclear power can be sustained while still deriving some net energy from the system? The comparison is not easy, because the energy involved in construction cannot be unambiguously determined and, further, the investment is primarily in the form of heat energy, which is less valuable than electrical energy. (Remember the energy conversion efficiency of about one-third.)

If we ignore the question of timing and growth, the answer is easy: the construction energy investment appears to be within a factor of two or three of 1% of the total lifetime output from the nuclear power plant. Even considering requirements for other facilities, the total lifetime energy investment in construction and fuel enrichment cannot be much greater than 6% or 7% of the system output. (The one imponderable that could raise this percentage would be the need to begin using the very low-grade uranium resources mentioned in Chapter 9. It is likely, though, that the nuclear power system would take a form that made more efficient use of uranium resources, rather than using large amounts of very low-grade ores.) And, even considering the fact that an initial core, with its associated enrichment, must be provided, the initial investment will not exceed about 3% of the lifetime output. Presuming that a plant operates about 30 years, this amount would be provided by about one year's average output. The time dependence of the net energy from a nuclear power plant is illustrated in Figure 2-5. This dependence indicates that the overall nuclear power system would have to grow extremely rapidly to result in a net energy loss. Since an operating plant can provide enough energy to start a new plant in about one year's operation, the system could, from a net energy point of view, roughly double in size each year. This is an unlikely prospect.

It is also important to note that other energy systems require a substantial energy investment (or at least carrying charge). Even fossil fuel power plants are large, complex facilities, particularly with the emission control systems that are now being required. The energy investments in such plants are less than the comparable investment in a nuclear power plant, but not altogether negligible. As for the energy required for fuel preparation, it is illuminating to contemplate the fact that, in present oil refineries, about 10% of each barrel of oil supplied to the plant is required for the plant's operation. The energy investment in nuclear power operations does not appear to be an overwhelming problem. There are other matters of much greater significance, as noted in the succeeding chapters.

ECONOMICS AND PRACTICALITY

A primary consideration in the choice and design of a reactor system is the total cost of the electricity it produces, including the cost of capital facilities, the cost of operations, and the cost of the fuel materials. For nuclear power, as for other systems, costs are changing rapidly and are highly controversial, especially when comparisons between systems are made. Because of the ill-defined nature of electrical generating costs, particularly as they may be projected for the future, we do not address these questions in any detail. Operators of nuclear power plants typically claim nuclear generated electricity is now cheaper than other types. But this situation may have changed, as the time it takes to license a power plant has increased. Moreover, the apparent costs may not include associated costs, such as

Figure 2-5. NET ENERGY FROM A NUCLEAR POWER PLANT.
The net energy derived from a nuclear power plant is indicated schematically as a function of the year of plant operation. The net energy is the output energy less the energy required to construct the plant and to build and operate associated fuel cycle facilities. Time zero marks the beginning of plant operation. After about two years of operation, the plant has produced enough energy to have paid back its own investment and start up a new plant.

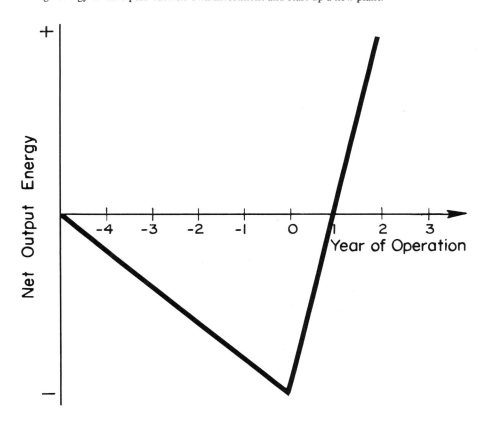

that of waste management. On the other hand, the same may be true of other types of power plants, such as those fired by coal. In any case, it is evident that these questions cannot be treated both briefly and adequately.

An associated question arises in connection with consideration of new reactor types: speaking practically, can they be commercialized by the time we might want them? The reactors described in Part II have already been on the market, but the development time for a new type is substantial, and building up the industrial base for manufacture takes even longer. For this reason, reactor types that are mild extensions of present technology can reach the market relatively soon. Radical departures, on the other hand, would have to wait their time. We have not treated commercialization time explicitly, although, in discussing the question of nuclear power growth (Chapters 9 and 10), some allowance for lead times has been implicit.

Bibliography — Chapter Two

"NTIS" indicates report is available from: National Technical Information Service, U.S. Department of Commerce, 5285 Port Royal Road, Springfield, VA 22161.

APS-1977. L. C. Hebel et al., "Report to the American Physical Society by the Study Group on Nuclear Fuel Cycles and Waste Management" (July 1977). To be published as a supplement to *Reviews of Modern Physics,* vol. 50.

General introduction to fuel cycles; treats implications of fuel reprocessing and waste management for various fuel cycles.

CONF-76-0701. "Proceedings of the International Symposium on the Management of Wastes from the LWR Fuel Cycle," Denver, July 11-16, 1976 (NTIS).

Many papers on various aspects of the management of LWR wastes.

Dahlberg, R. C., Turner, R. F., and Goeddel, W. V., "HTGR Fuel and Fuel Cycle Summary Description," General Atomic Company report GA-A12801 (rev., January 1974).

Describes HTGR fuel cycle, as well as the economics and uranium requirements of the HTGR.

ERDA-1. "Report of the Liquid Metal Fast Breeder Reactor Program Review Group," U.S. ERDA report, ERDA-1 (January 1975) (NTIS).

Presents uranium requirements for many reactor types, particularly in appendixes.

ERDA-76-162. "The Management and Storage of Commercial Power Reactor Wastes," U.S. ERDA report ERDA-76-162 (1976) (NTIS).

Brief summary of the need for and possible methods of managing LWR wastes.

Foster, J. S., and Critoph, E., "The Status of the Canadian Nuclear Power Program and Possible Future Strategies," *Annals of Nuclear Energy,* vol. 2, p. 689 (1975).

Discusses features of the CANDU family of reactors, including their economics and uranium requirements.

GESMO. "Final Generic Environmental Statement on the Use of Recycle Plutonium in Mixed Oxide Fuel in Light-Water Cooled Reactors: Health, Safety, and Environment," 5 vols., U.S. NRC report NUREG-0002 (August 1976) (NTIS).

Detailed treatment of the environmental considerations associated with the recycle of plutonium from reprocessed LWR fuel.

NUREG-0001. "Nuclear Energy Center Site Survey — 1975," 5 vols., U.S. NRC report NUREG-0001 (January 1976) (NTIS).

A survey for sites in the United States that might be suitable for "energy centers," assemblies of several nuclear facilities.

NUREG-0116. "Environmental Survey of the Reprocessing and Waste Management Portions of the LWR Fuel Cycle" (supplement 1 to WASH-1248). U.S. NRC report NUREG-0116 (October 1976) (NTIS).

Describes environmental impacts of LWR fuel reprocessing and waste management.

ORNL-4451. "Siting of Fuel Reprocessing Plants and Waste Management Facilities," Oak Ridge National Laboratory report ORNL-4451 (July 1971) (NTIS).

An older standard reference treating the effect of fuel reprocessing and waste management facilities on local and worldwide concentrations of radionuclides.

Pigford, T. H., et al.: "Fuel Cycles for Electric Power Generation," Teknekron report EEED 101 (January 1973, rev. March 1975); "Fuel Cycle for 1000-MW Uranium-Plutonium Fueled Water Reactor," Teknekron report EEED 104 (March 1975); "Fuel Cycle for 1000-MW High-Temperature Gas-Cooled Reactor," Teknekron report EEED 105, (March 1975). These are included in "Comprehensive Standards: The Power Generation Case," U.S. EPA report PB-259-876 (March 1975) (NTIS).

Presents material quantities and environmental effluents in the fuel cycles for various technologies, including nuclear power.

Pigford, T. H., and Ang, K. P., "The Plutonium Fuel Cycles," *Health Physics,* vol. 29, p. 451 (1975).

Calculates quantities of plutonium and other heavy elements processed in the fuel cycles of several types of reactors.

WASH-1174(74). "The Nuclear Industry," U.S. AEC report WASH-1174(74) (1975) (NTIS).

Summarizes the state of the commercial nuclear industry as of late 1974; specifies the operator and location of various fuel cycle facilities.

WASH-1209. "The Potential Radiological Implications of Nuclear Facilities in a Large Region of the United States in the Year 2000," U.S. AEC report WASH-1209 (January 1973) (NTIS).

Estimates the size and impact of the radioactive emissions of a large nuclear power system.

WASH-1248. "Environmental Survey of the Uranium Fuel Cycle," U.S. AEC report WASH-1248 (April 1974) (NTIS).

Summary of environmental impacts of the fuel cycle for light-water reactors.

WASH-1250. "The Safety of Nuclear Power Reactors (Light-Water Cooled) and Related Facilities," U.S. AEC report WASH-1250 (July 1973) (NTIS).

Gives some basic information on light-water reactors and on their fuel cycles.

Wilson, R., and Jones, W. J., *Energy, Ecology, and the Environment*, (Academic, New York, 1974).

Discusses many of the "environmental" aspects of energy technologies.

CHAPTER THREE

Nuclear Power Plant Emissions

THE RADIOACTIVE MATERIALS produced in the course of the nuclear chain reaction are the primary reason for much of the cost and complexity of nuclear power plants, as well as for the great emotions that their use has engendered. We have given some attention to the nuclear processes of main concern in a reactor: fission and conversion. However, we have not yet described the full array of radionuclides present in an operating nuclear power plant. Many of these radionuclides may emit radiation that is harmful to human beings. Few of them, however, can reach the environment in significant quantities under ordinary operating conditions. Regulations now applicable to nuclear power plants require incorporation of radioactive emission control systems that have an effectiveness directly tied to the current understanding of the harmful effects of radiation. The result is that a nuclear power plant ordinarily contributes a radiation dose to individuals in its vicinity that is well below the dose they receive from natural radiation sources. This chapter briefly describes the nature and effect of routine emissions from nuclear power plants. The risk from accidents at these plants will not be examined until the next chapter.

RADIOACTIVE INVENTORY

Most of the radioactivity present in an operating nuclear reactor arises from the actual operation of that reactor. Only a small portion of a reactor's radioactive inventory, i.e., the amount of radioactivity present in the reactor, is introduced into the reactor in the form of fuel. The remainder is produced by the nuclear chain reaction and related processes. Some care must be taken in measuring "radioactivity." A nucleus is radioactive if it has the capability of "decaying" to another nucleus, emitting some sort of radiation in the process (see Figure 3-1). Radioactivity is measured in terms of the rate at which such decays occur. A quantity of materials in which 3.7×10^{10} decays (or "disintegrations") occur per second is said

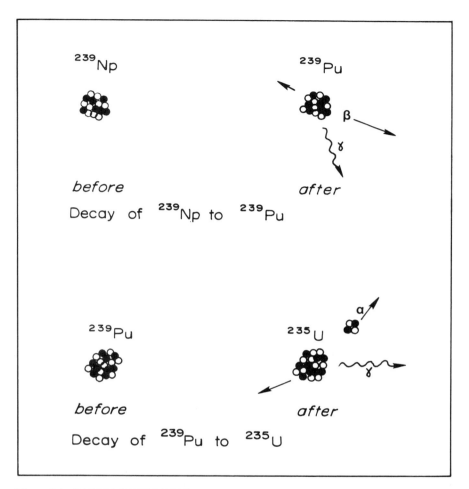

Figure 3-1. RADIOACTIVE DECAY.
Certain nuclides may "decay" to other nuclides, in the process emitting radiation. The upper part of the figure illustrates the beta decay of neptunium 239 to plutonium 239. The lower part shows alpha decay of plutonium 239; as shown in the drawing, an alpha particle consists of two protons and two neutrons, the basic components of all nuclei. In neither example is the decay stimulated by an external agent. Decay occurs spontaneously, with a probability per unit time that is inversely proportional to the half-life of the radionuclide. The half-life of ^{239}Np is two days; that of ^{239}Pu is 24,000 years. The electrons surrounding the nuclei are not shown. In both cases, the decay may be accompanied by the emission of a gamma ray by the residual nucleus, as indicated. The decay products all have energy that is absorbed through collisions with the surrounding material.

to constitute one Curie (abbreviated Ci) of radioactivity. This unit indicates only the decay rate, not the type of radiation given off or the possible effect on humans.

The radioactive inventory of a nuclear power plant arises from several sources. A relatively small contribution is made from natural radioactivity extracted from the earth. In a uranium-fueled reactor, this contribution — largely in the form

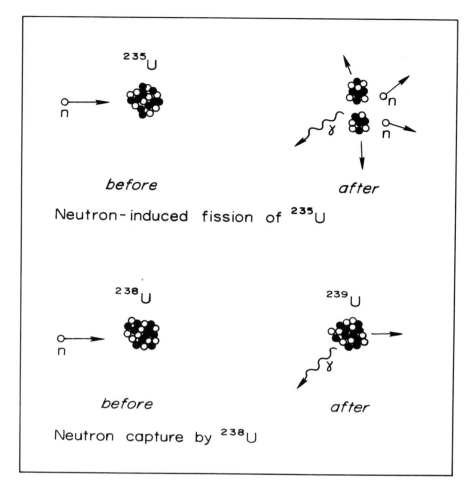

Figure 3-2. NUCLEAR REACTIONS.
The figure illustrates two important types of nuclear reactions. The upper is the fission of uranium 235, the reaction that liberates most of the energy in a uranium-fueled light-water reactor. The lower illustrates the capture of a neutron by uranium 238, the first step in the conversion of uranium 238 to plutonium 239.

of uranium isotopes, 234, 235, and 238 — is a few hundred Ci. In a reactor that recycles plutonium or uranium, the radioactivity contained in the fresh fuel may be considerably higher. It is still much less than the radioactivity produced by reactor operation.

Two major modes of radioactivity production are associated with the fuel itself. Fission produces fragments that are radioactive, and nonfission reactions, such as neutron capture, produce heavy radioactive nuclei (see Figure 3-2). A third major mode of radioactivity production is the activation of cladding and structural materials.

The fissionable nuclei of interest have masses in the range 230 to 245. The products resulting from fission of a given nuclide, such as uranium 235, vary from one fission to another. Generally, two major fragments are produced, of almost but not quite equal mass. If many fissions are observed, the distribution of fission fragment masses is found to have peaks in the vicinity of masses 95 and 140. The most important examples of the first group are krypton and strontium, and of the second, iodine, xenon, and cesium. Another important product of fission, sometimes given off as a third fission fragment, is tritium, the mass 3 isotope of hydrogen.

The second radioactivity production mode associated with the reactor fuel is nonfission reactions which transform fuel nuclides into other nuclides of similar mass. These nuclides, many of which are radioactive, belong to the "actinides," a chemical classification that includes the elements thorium, protactinium, uranium, neptunium, plutonium, americium, and curium. As we have noted, thorium and uranium occur naturally in significant quantities. Of the actinides that are produced in a reactor, the bulk arise from neutron capture by fuel materials, possibly followed by subsequent decays. We have referred to this process as "conversion" when the result is to convert a nonfissile nucleus to one that is fissile.

Radioactivity may also be produced in material other than fuel. The most important source of this type is, in fact, the fuel cladding and fuel assembly structural materials. These materials may become "activated" by reactions similar to those that produce the actinides. Smaller amounts of radioactivity are produced in other reactor materials, including the control rods, the coolant, and the reactor structure itself. This radioactivity is produced primarily by the interaction of neutrons with these materials.

Table 3-1 presents information on radioactivity in light-water reactor power plants, emphasizing the radionuclides that may be important from the point of view of possible releases. The table gives the radionuclides, their half-lives (the time necessary for half of an initial amount of a radionuclide to decay), the amount in an operating reactor (if all the fuel were at discharge composition), the amount in the yearly fuel discharge (for various times after discharge), physical characteristics, and health considerations. The radionuclides are grouped according to their source. See Appendix A for a list of chemical symbols.

Note that the total amount of radioactivity in an operating reactor is about 15 billion Ci. This large inventory of radioactivity is the source of the heat that remains in a reactor even after it is shut down. Because many of these radionuclides have very short half-lives, their contribution to the radioactive inventory diminishes rapidly after shutdown. The result is that the total inventory decreases rapidly. For example, the yearly fuel discharge contains about 5 billion Ci at shutdown, but this decreases to about 0.1 billion Ci within 150 days.

The above discussion has emphasized radioactivity in the fuel or its structural materials. A nuclide that is produced both in the core and coolant and that has received most attention recently is carbon 14. As measured in Curies, this isotope is produced in much smaller amounts than many of the other isotopes. However, it is more difficult to control and has a much longer half-life than many of the others,

TABLE 3-1.
Radioactivity in Reactor and Fuel Cycle for a 1000-MWe Uranium-Fueled Light-Water Reactor[a]

Radionuclides		Half-Life	Reactor Inventory[b] (10⁶ Curies)	In Discharge Fuel[c] (10⁶ Curies/yr)			Elemental Boiling Temperature (°F)	Health Considerations and Principal Radiation
				At Discharge	150-Day Decay	10-Yr Decay		
Fission Products								
Tritium[d]	^{3}H	12.26 yr	0.0723	0.0241	0.0239	0.0139	212°F (as HTO)	Internal hazard, β (0.019 MeV max)
Krypton	83m	1.86 hr	5.71	1.90	0	0	−243°F	External irradiation, β (0.67 MeV max for ^{85}Kr)
	85m	4.4 hr	17.2	5.70	0	0		
	85	10.76 yr	1.16	0.383	0.373	0.201	(Gaseous)	
	87	76 min	34.0	11.3	0	0		
	88	2.8 hr	49.0	16.5	0	0		
	89	3.18 min	61.8	20.3	0	0		
	90	33 sec	58.5	22.7	0	0		
Total[e]			325	107.7	0.373	0.201		
Strontium	89	52.7 days	71.6	23.8	3.22	0	2,490°F Nonvolatile	Internal hazard to bone and lung, β (0.55 MeV max for ^{90}Sr)
	90	27.7 yr	7.80	2.58	2.56	2.02		
Total[e]			526	174	5.78	2.02		
Iodine	129	1.7×10⁷ yr	3.03×10⁻⁶	1.00×10⁻⁶	1.02×10⁻⁶	1.03×10⁻⁶	333°F	Internal hazard to thyroid, β (0.8 MeV max for ^{131}I) and γ
	131	8.05 days	71.9	23.9	6.01×10⁻⁵	0		
	132	2.26 hr	103	34.2	0	0	(Highly volatile)	
	133	20.3 hr	137	45.6	0	0		
	134	52.2 min	156	51.8	0	0		
	135	6.68 hr	123	40.7	0	0		
	136	83 sec	54.0	17.9	0	0		
Total[e]			1,017	337	6.11×10⁻⁵	1.03×10⁻⁶		
Xenon	131m	11.8 days	0.582	0.193	9.05×10⁻⁵	0	−162°F (Gaseous)	External irradiation, β (0.35 MeV max for ^{133}Xe) and γ
	133m	2.26 days	3.29	1.09	0	0		
	133	5.27 days	137	45.6	0	0		
	135m	15.6 min	36.8	12.2	0	0		
	135	9.14 hr	25.7	8.52	0	0		
	137	3.9 min	132	43.8	0	0		
	138	17.5 min	128	42.5	0	0		
	139	43 sec	107	35.6	0	0		
Total[e]			680	225	9.05×10⁻⁵	0		
Cesium	134	2.046 yr	19.0	6.32	5.49	0.215	1238°F (Volatile)	Internal hazard to muscle, β (1.176 MeV max for ^{137}Cs) and γ
	137	30.0 yr	9.92	3.29	3.26	2.61		
Total[e]			595	198	8.75	2.83		
Zr, Nb, Mo, Tc, Ru, Rh, Pd, Ag, Cd, In, Sn, Sb			1,880	625	54.1	0.0450	Nonvolatile	External and internal hazard
Rare Earths: La, Ce, Pr, Nd, Pm, Sm, Eu, Gd, Tb, Dy, Ho			4,140	1,374	48.8	0.416	Nonvolatile	External and internal hazard
Total fission products			11,970	3970 (1,067 Te/yr)	130	9.98		

TABLE 3-1.
Radioactivity in Reactor and Fuel Cycle for a 1000-MWe Uranium-Fueled Light-Water Reactor[a] (continued)

Radionuclides	Half-Life	Reactor Inventory[b] (10^6 Curies)	In Discharge Fuel[c] (10^6 Curies/yr)			Elemental Boiling Temperature (°F)	Health Considerations and Principal Radiation
			At Discharge	150-Day Decay	10-Yr Decay		
Actinides							
Uranium 237	6.75 day	108	35.9	7.42×10^{-6}	0	Nonvolatile	External and internal hazard, β (1.29 MeV max for ^{237}U) and γ
239	23.5 min	1,708	566	0	0		
Total[e]		1,816 (99.7 Te)	601 (33.0 Te/yr)	2.72×10^{-5}	2.15×10^{-5}		
Plutonium 238	86.4 yr	0.138	0.0459	0.0483	0.0463	Nonvolatile	Internal hazard to bone, liver, lung, lymph, α (about 5 MeV for ^{238,239,240,242}Pu) and β (for ^{241}Pu)
239	24,390 yr	0.0318	0.0105	0.0107	0.0107		
240	6,580 yr	0.0500	0.0166	0.0166	0.0166		
241	13.2 yr	12.4	4.11	4.04	2.42		
242	3.79×10^5 yr	1.24×10^{-4}	4.14×10^{-5}	4.14×10^{-5}	4.14×10^{-5}		
243	4.98 yr	22.2	7.35	0	0		
Total[e]		34.9 (0.893 Te)	11.57 (0.296 Te/yr)	4.11	2.49		
Fissionable Pu (239Pu + 241Pu)		(0.627 Te)	(0.207 Te/yr)				
Americium (241, 242m, 242, 243) and Curium (242, 243, 244)		1.14	3.79	0.340	0.0611	Nonvolatile	Internal hazard, α (5.5 MeV for ^{241}Am)
Total Actinides Th, Pa, U, Np, Pu, Am, Cm		3,614	1,198	4.45	2.55		
Cladding							
Activated Zircaloy cladding and Inconel spacers (Cr, Mn, Fe, Co, Ni, Zr, Nb, Sb)		12.9 (28.2 Te)	4.28 (9.36 Te/yr)	0.967	0.109	Nonvolatile	
TOTAL FISSION PRODUCTS, ACTINIDES, AND CLADDING[f]		15,600 (132 Te)	5,170 (43.9 Te/yr)	135	12.6		

a. Overall thermal efficiency = 32%.
b. Calculated on the basis that all fuel has the composition of discharge fuel. For nonsaturable species, whose half-lives are large compared with the typical fuel life of 3 years, the inventory within the reactor will be about one-half to two-thirds of the value listed, depending upon the fuel management scheme. The inventories listed in the table are accurate for saturable (short-half-life) species. Material quantities in metric tonnes (Te) are listed in parentheses.
c. Fuel cycle quantities are calculated on the basis of a fuel exposure of 33,000 MWd/Te 100% load factor. Material quantities are listed in parentheses.
d. Quantities listed are for fission-product tritium only and do not include tritium formed by neutron-activation reactions in boron controls and contaminants.
e. Totals include all isotopes of the element, including short-lived species not listed in the table.
f. Does not include 14C, which is produced in the fuel and cladding (the order of 50 Ci/yr), as well as the coolant (roughly 10 Ci/yr).

Source: Adapted from T. H. Pigford et al., "Fuel Cycles for Electric Power Generation," Teknekron report EEED 101 (January 1973, rev. March 1975).

so that it may be regarded as more significant. This matter is discussed more fully in the next section.

RADIATION EFFECTS AND RADIOLOGICAL STANDARDS

The radioactivity contained in a nuclear power plant poses a threat to humans primarily through its capacity for emitting radiation. This radiation may take a number of forms, including energetic particles such as beta rays (actually electrons or positrons) and alpha particles (helium nuclei) and electromagnetic radiation such as gamma and X-rays. These forms share the feature of being energetic enough to cause the formation of ion pairs in material through which they pass. Such ionization is the mechanism whereby this radiation may harm living organisms.

In passing through tissue, radiation deposits energy in that tissue, primarily through the ionization mechanism just mentioned. The fundamental measure of radiation dose is the amount of energy deposited per unit mass of material. The unit normally used is the "rad," equal to 100 ergs deposited per gram of tissue. However, the energy deposited is itself not a measure of the biological damage caused by the radiation. This damage depends on many factors, including the size and distribution of the dose, the rate at which it is administered, and also the radiation being considered, including both its type and its energy. The question of dose size and dose rate is considered briefly here and in the next chapter. The biological effectiveness of the radiation is often related to the rate at which its energy is deposited along its path through tissue. For small doses and low dose rates, it has been convenient to introduce another quantity, called the "dose equivalent," which is actually the dose (given in rad) multiplied by a factor that indicates the relative biological damage due to the radiation being considered. The unit of dose equivalent is the "rem." For certain types of radiation, such as X-rays, the dose in rad and the dose equivalent in rem are numerically equal. For other types, such as alpha particles, with their higher effectiveness for damaging tissues, the dose equivalent (in rem) may be higher by as much as a factor of 20 than the dose (in rad). As a final specification, the region of the body that is exposed must be given; however, in much of this discussion, it is presumed that the whole body is exposed.

The effects on humans may be divided into two categories: acute and long-term. The passage of radiation causes damage at the cellular level that, depending on the dose size and dose rate, may show itself almost immediately or after a long latency period. Whole body doses greater than 25 rad, delivered over a short period (such as minutes or hours), can result in sickness, with death becoming increasingly probable as the dose is increased; by 1000 rad, death is virtually certain within a few weeks. Members of the general public (and, indeed, radiation workers) can be exposed to doses in this range only under very unusual circumstances: with the use of nuclear weapons or after an extremely serious nuclear accident.

At smaller doses, or even similar doses spread over a long period of time, the damage is not sufficient to cause radiation sickness or death. However, the damage has the potential for leading to effects that show themselves years later. Although a number of such effects may be considered, two are of greatest concern: cancer, which directly affects the exposed individuals, and genetic damage, which affects succeeding generations. In either case, the probability of an effect — whether a cancer or a genetic defect — may be related to the size of the dose (or, more appropriately, dose equivalent), the conditions of exposure, and the type of individual exposed. We cannot treat these questions in any detail. As discussed later in this chapter (and in the next), it is often supposed, for regulatory or assessment purposes, that the probability of radiation effects at low doses and dose rates is directly proportional (i.e., linearly related) to the size of the dose. This results in the great simplification that a given radiation dose is presumed to have the same probability of producing an effect, regardless of how it is distributed among a population group. (Thus the dose may be to one individual or may be the sum of doses to a number of individuals. However, this presumes that none of the individuals considered receives a dose sufficient to cause acute effects.) The population dose is often expressed in "person-rem." Presuming the linear dose-response just mentioned, about 5000 to 20,000 person-rem (i.e., the exposure of a large number of individuals to a total of 5000 to 20,000 rem) is often taken to yield, on the average, one cancer death.

Generally, radiation protection standards have not been based on this presumption of linearity. The greatest attention, in the early formulation of standards, was given to individuals who are occupationally exposed. For such situations, protection standards are often set about a factor of 10 below exposure levels that have been observed to cause harmful effects. Standards for members of the general public are often set another factor of 10 lower. This philosophy affords a high degree of protection where an exposure "threshold" exists, i.e., where no harmful effects occur below some minimum exposure level. Where such a threshold may not exist, exposure limiting standards afford a degree of protection that can be determined only on the basis of a dose-response function, such as the linear function mentioned above.

In any case, the basic radiation protection criteria recommended by a number of national and international bodies are as follows: that those who are occupationally exposed should receive no more than 5 rem/year in the course of their work; that individual members of the general public should receive no more than 0.5 rem/year; and that large population groups should receive no more than 0.17 rem/year. (The last limit is based on consideration of the potential for genetic damage.) In every case, these are limits on exposures from human sources, but excluding medical practice.

These are limits on dose equivalent, not on environmental concentrations of radionuclides or on power plant emissions. Maximum permissible concentrations may be derived from the dose limiting recommendations. This derivation requires knowledge of the manner in which radionuclides can expose the body to radiation, a matter of great complexity in many cases. Examples of concentration limits of interest to nuclear power operations are given in Table 3-2. The limits are given for air and water, the two main fluids that humans ingest. The occupational limits presume individuals are exposed to these concentrations for 40 hours per week; the general public limits presume 168 hours per week. For many radionuclides, the

TABLE 3-2.

Maximum Permissible Concentrations of Selected Nuclear Power Related Radionuclides[a]

$(\mu Ci/m\ell)$

Element (Atomic Number)	Isotope	Form[b]	Occupationally Exposed (40 hours/wk) Air	Water	General Public (168 Hours/wk) Air	Water
Carbon (6)	14	S	4×10^{-6}	2×10^{-2}	1×10^{-7}	8×10^{-4}
		CO_2	5×10^{-5}	–	1×10^{-6}	–
Cesium (55)	137	S	6×10^{-8}	4×10^{-4}	2×10^{-10}	2×10^{-5}
		I	1×10^{-8}	1×10^{-3}	5×10^{-10}	4×10^{-5}
Iodine (53)	131	S	9×10^{-9}	6×10^{-5}	1×10^{-10}	3×10^{-7}
		I	3×10^{-7}	2×10^{-3}	1×10^{-8}	6×10^{-5}
Krypton (36)	85		1×10^{-5}	–	3×10^{-7}	–
Plutonium (94)	239	S	2×10^{-12}	1×10^{-4}	6×10^{-14}	5×10^{-6}
		I	4×10^{-11}	8×10^{-4}	1×10^{-12}	3×10^{-5}
Radon (86)	222		3×10^{-8}	–	1×10^{-9}	–
Strontium (38)	90	S	1×10^{-9}	1×10^{-5}	3×10^{-11}	3×10^{-7}
		I	5×10^{-9}	1×10^{-3}	2×10^{-10}	4×10^{-5}
Uranium (92)	235	S	5×10^{-10}	8×10^{-4}	2×10^{-11}	3×10^{-5}
		I	1×10^{-10}	8×10^{-4}	4×10^{-12}	3×10^{-5}

a. Abstracted from 10 CFR 20.
b. The form of the radionuclide is usually specified by a letter, S for soluble, I for insoluble.
Source: A. V. Nero and Y. C. Wong, "Radiological Health and Related Standards for Nuclear Power Plants," Lawrence Berkeley Laboratory report LBL-5285 (January 1977).

limits are very small indeed, considering the large amounts of these radionuclides inside a nuclear power plant. However, this inventory is so well controlled that, under normal operating conditions, these limits are never approached.

Often, in fact, the nuclear-power-caused alteration of radiation levels is difficult to measure because of natural background contributions. Cosmic rays and radionuclides in the earth, the air, and even our bodies contribute an average dose of about 0.10 rem/year, and this dose may vary about 50% from place to place. This variation is so much larger than ordinary nuclear power plant contributions that great care must be taken to measure those contributions. Finally, it is worth noting that medical exposures amount, on the average, to about 0.07 rem/year, somewhat smaller than the natural background exposure rate, but still much larger than the average contribution expected from nuclear power operations, on the order of 0.001 rem/year.

TYPICAL EFFLUENTS FROM NUCLEAR POWER PLANTS

As suggested in Chapter 1, natural barriers serve to limit the release of much of the radioactive inventory of a nuclear power plant. Only the radionuclides that are gaseous or volatile or that are produced outside the fuel rods can routinely escape to the environment in significant quantities. Such radionuclides were noted above, particularly in Table 3-1. The most important routine releases are tritium, the noble gases krypton and xenon, iodine, and carbon. All but the last are produced during fission and, because of their chemical characteristics, some of the reactor's inventory can escape — often due to fuel rod failure — into the environment. Carbon 14, on the other hand, is produced largely by neutron interactions with ^{14}N and ^{17}O in the fuel and coolant, and that produced in the coolant can escape from the power plant. Typical amounts escaping yearly from a 1000-MWe light-water reactor power plant are (in Curies):

| ^3H(tritium) | 100 to 1000 | Kr, Xe | 300 to 50,000 |
| $^{129,131}I$ | 0.01 to 0.03 | ^{14}C | about 8 |

These are emitted into the air, with the exception of tritium, a large portion of which may be discharged into water. These emissions represent only a portion of the total amount of these radionuclides produced in a reactor. The bulk is released after the fuel rods are chopped up and dissolved at a reprocessing plant.

The amount of effluent depends strongly on the control measures incorporated into a particular plant. Control systems may be divided into two functional categories: filtration and holdup. Filtration systems do exactly what the word implies; they physically or chemically separate the radionuclides of interest from the plant's gaseous and liquid waste streams. Holdup systems take advantage of the fact that some of the radionuclides of interest have relatively short half-lives; these systems contain the waste streams for a period long enough to permit a substantial portion of the radioactivity to die off, after which discharge to the environment may take place. The plant gaseous and liquid waste handling systems may be quite complex. Figures 3-3 and 3-4 indicate these systems schematically for a pressurized-water reactor.

The details of these control systems vary from one plant to another and certainly from one plant type to another. However, the principles of operation do not vary drastically. As noted in the next section, the degree to which control systems have to be incorporated in the design of a particular plant is determined on the basis of a calculation of the benefit to be derived from the resulting limitation of exposure of the population surrounding that plant.

Overleaf: Figure 3-3. NUCLEAR PLANT GASEOUS WASTE DISPOSAL AND VENTILATION SYSTEM.
A nuclear plant provides for handling of various types of gaseous waste and for ventilation of the plant buildings. In the system shown the high-activity gas wastes, including those from the makeup and purification system (see Figure 3-4), are subjected to charcoal adsorption, then stored in tanks, after which they may be vented to the atmosphere or recycled. Other gaseous wastes are vented after filtration and adsorption. (Figure courtesy of U.S. NRC.)

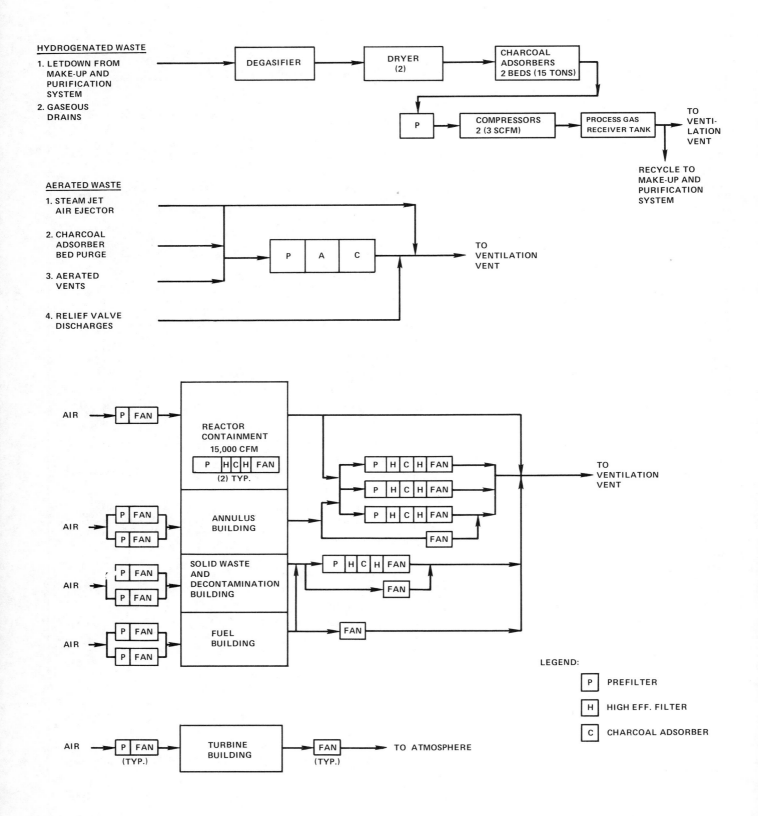

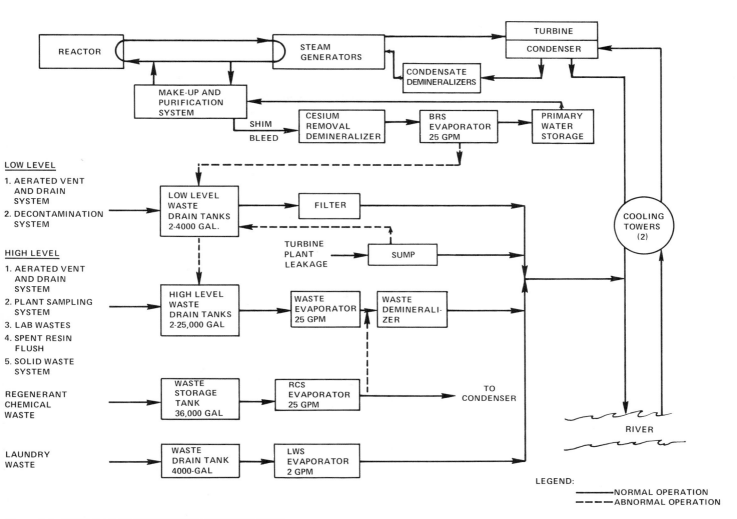

Figure 3-4. NUCLEAR PLANT LIQUID WASTE SYSTEM.
The system for handling and cleaning liquids in a nuclear power plant is very complex. In a typical plant, water is used for cooling, heat transfer, and miscellaneous purposes. Prior to release of any of this water to the general environment, it is filtered, demineralized, or held up. (Figure courtesy of U.S. NRC.)

KEEPING EXPOSURES "AS LOW AS IS REASONABLY ACHIEVABLE"

A large number of possible exposure modes must be considered in calculating the dose to humans from operation of a nuclear power plant. Many of these modes are indicated in Figure 3-5. Ultimately, exposures may arise from three sources: gaseous effluents, liquid effluents, and direct irradiation. The first two are the dominant modes by far, except when the transport of irradiated fuel is considered. The direct irradiation of members of the general public by radioactive materials that are still contained at the nuclear power plant is very small compared with irradiation from radionuclides that are released from the plant site.

Earlier in this chapter, we cited the dose-limiting recommendations that are

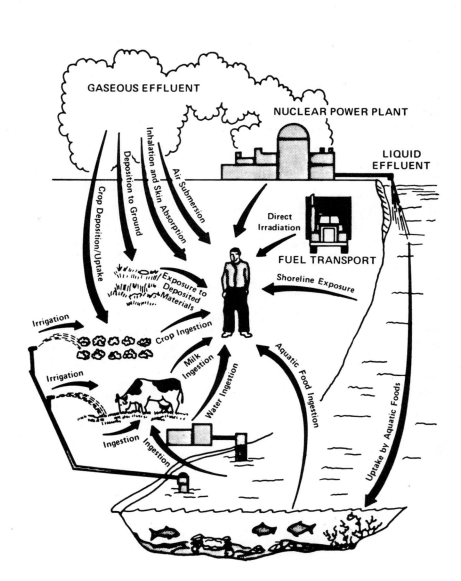

Figure 3-5. RADIATION EXPOSURE PATHWAYS TO HUMANS.
Humans may suffer radiation exposures from power plants in various ways. These include exposure to radiation from decay of radioactive material that is released from plants into the atmosphere or into water resources, as well as exposure to radiation coming directly from the power plant or, more likely, from transport of irradiated fuel.
(Figure reproduced from Koshkonong Draft Environmental Statement.)

TABLE 3-3.

Numerical Design Limits for Dose from Nuclear Power Plants to Members of the General Public

Origin and Type of Dose	Design Objective (per reactor)
Liquid effluents	
Dose to total body from all pathways	3 mrem/yr
Dose to any organ from all pathways	10 mrem/yr
Gaseous effluents (only for noble gases)	
Gamma dose in air	10 mrad/yr
Beta dose in air	20 mrad/yr
Dose to total body	5 mrem/yr
Dose to skin	15 mrem/yr
Radioiodines and particulates released to the atmosphere	
Dose to any organ from all pathways	15 mrem/yr

Source: 10 CFR 50 Appendix I (see text).

applicable to all human activities. Another general prescription is that exposures be kept "as low as is reasonably achievable," a prescription that has recently superseded the older "as low as is practicable" criterion. As applied to nuclear power plants, "as low as is reasonably achievable" has led to very practical requirements in the United States. These requirements are embodied in Appendix I to Part 50 of the Code of Federal Regulations, Title 10; this title sets the powers and responsibilities of the Nuclear Regulatory Commission (NRC) in regulating nuclear power operations. Appendix I requires that light-water reactor power plants licensed for commercial operation meet certain numerical design objectives for maximum exposures of individual members of the general public. These maximum exposures are given in Table 3-3. The units used are mrad and mrem, which are, respectively, 0.001 rad and 0.001 rem. The design objectives are 3 and 5 mrem/year for whole body dose from liquid and gaseous effluents, respectively, with somewhat larger doses permitted for specific organs. These whole body doses are to be compared with the generally applicable dose limit of 500 mrem/year (to members of the general public) and with the natural background dose of about 100 mrem/year.

As it turns out, nuclear power plants that are now being licensed meet these objectives. The maximum whole body dose from either exposure path (liquid or gaseous) is typically about 1 mrem/year. The average dose to any large population group surrounding a nuclear power plant is much less than this.

Appendix I also requires that additional control measures be implemented until the resulting reduction in population exposures is no longer cost effective. The tentative criterion applies to doses within 50 miles of the plant and specifies that a reduction of population exposure by one person-rem shall be valued at $1000. If a linear dose-response relationship is assumed, and 5000 person-rem is taken to be equivalent to one death, this effectively values a life at $5 million, a figure that is considerably higher than that used in many considerations (such as the question of whether sulfur control equipment should be required on coal-fired power plants).

In any case, for the nuclear power plants now being licensed, the population dose is relatively small. A typical value is about 5 person-rem/year to the population within 50 miles. Somewhat surprisingly, about half of this dose is contributed by irradiation of these populations by the transport of spent fuel. These values depend strongly on the particular site (see, for example, Figure 3-6). However, as discussed briefly below, the overall population dose from other fuel cycle facilities is considerably larger than that from the nuclear power plants themselves.

Before proceeding, it is worth remarking that the 5 person-rem/year just noted is approximately the dose to which a single radiation worker may be exposed, according to the dose limiting recommendation discussed earlier in this chapter. In fact, the regulations of the NRC also require that doses to these workers be as low as practicable. In practice, this turns out to result in a total yearly dose of

Figure 3-6. FLOATING NUCLEAR POWER PLANT.
The manner in which nuclear plants interact with the environment depends heavily on their sites. Proposed floating nuclear power plants would reside in the ocean, therefore having exposure pathways significantly different from those of ordinary plants. This would have particular significance in the event of a major nuclear plant accident; for an ordinary LWR, a molten core would come to rest in the earth beneath the plant (see next chapter); for a floating plant, a body of water would be in immediate danger of contamination. The plant shown in the figure incorporates a PWR system and was designed for use on the Atlantic coast. (Figure courtesy of Offshore Power Systems.)

about 500 person-rem to the workers at a nuclear power plant. It is clear, then, that these workers accept the bulk of the risk, both individually and as a group, from routine operation of a nuclear power plant.

Most of the dose to workers is from direct irradiation, whereas the dose to members of the general public is from radionuclide releases. These releases pose a difficulty in analysis to which we have not yet alluded. This difficulty arises from the fact that, once released, the radionuclides, and their daughters, may remain available to cause exposures for many years and even millenia. The population doses that we noted above (typically about 5 person-rem/year) include only the doses during the period when the plant is operating. If we consider certain long-lived radionuclides, such as carbon 14, with its half-life of 5700 years, the cumulative population dose over the millenia after their release dominates the dose from the shorter-lived radionuclides, even though it is the latter nuclides that dominate the dose rate in our lifetimes. It is difficult to consider very small dose rates lasting millenia on the same basis as we consider much higher dose rates that occur today. Moreover, these lower dose rates are many orders of magnitude lower than the dose rates from natural background radiation and even from the naturally occurring concentrations of the same radionuclides (such as ^{14}C). However, this long-term "dose commitment" must be considered if a strict assessment of population dose is to be made in the context of a cost-benefit comparison.

THE BALANCE OF THE FUEL CYCLE

We do not intend to treat in any detail the contributions of the rest of the fuel cycle to radiation doses to either workers or the public. However, for purposes of comparison, it is useful to note where other exposures arise and how they compare with those from the nuclear power plant. The basic fuel cycle functions and their environmental releases are discussed in Appendix F.

The principal sites for routine releases, aside from power plants, are extraction operations (mining and milling) and fuel reprocessing plants. The releases at extraction are primarily naturally occurring daughters of uranium and thorium. The releases at reprocessing, which is not now taking place commercially, would be similar to, but larger than, the releases from a nuclear power plant.

Whether the exposures from these sites are calculated to exceed those from a nuclear power plant depends on the method used to analyze the long-lived radionuclides. Uranium ore "tailings" piles, remnants from extraction operations, slowly generate and release radon, a radioactive gas whose decay daughters may contribute a large portion of the general public's routine exposure from nuclear power operations. Radon is generated over millenia from the decay of the long-lived isotopes of thorium and radium in the tailings. The exposure rate for populations downwind is immeasurably small. However, if the population dose is summed for long periods after the tailings are deposited, the total population dose becomes much larger than the few person-rem per reactor-year given earlier in this chapter. (In a similar way, the dose commitment from the power plant itself can become much larger if the carbon 14 emissions are considered over a long period.)

In a fuel reprocessing plant, the fuel rods are opened, and a large amount of radioactivity is liberated. Specific measures must be implemented if the more volatile species are to be controlled. Using the plant at Barnwell, South Carolina, as a model,[1] we may take the maximum annual dose to any members of the general public to be about 10 mrem/year, somewhat larger than for a nuclear power plant. Moreover, the most substantial emissions, those of tritium and the noble gases, are of sufficient quantity to alter the average radiation dose to humans through the world if reprocessing were introduced on a large scale. The alteration would be less than 1 mrem/year, but this could contribute the most significant population dose from nuclear power operations (depending, again, on how long-lived radionuclides are considered). However, introduction of noble gas control measures, as appears to be required by recently promulgated regulations of the Environmental Protection Agency, would reduce the maximum dose and the population dose substantially.

We have not discussed the impact of releases from waste disposal sites. The form of waste disposal has not yet been chosen, so that it is difficult to determine the associated contribution to human exposures. It does, however, appear unlikely that these contributions would exceed those from other fuel cycle facilities. (See Chapter 11 for further discussion of fuel reprocessing and waste disposal.)

In evaluating the importance of very slight alterations in the average population exposure, it is useful to keep in mind that these increases are much smaller than the local variations that humans experience in changing locations, or even living arrangements. In the case of mining and milling operations, which, depending on how dose commitments are calculated, may cause the largest population dose, the emissions are orders of magnitude smaller than natural emissions of the same radionuclides, simply because uranium and thorium occur naturally throughout the earth's crust. (These radionuclides are also released from coal-fired power plants.) The contributions of routine emissions from nuclear power operations are not to be ignored. However, it is difficult to make planning decisions based on this information, simply because the exposures are so small and comparable impacts of other human activites are not well understood.

Independently of this question, it is clear that exposures arising from routine releases from extraction operations and from fuel reprocessing can be as large as or larger than those from routine power plant emissions. However, when the effect of accidents is considered, this balance may change. The potential effects of accidents at nuclear power plants appear to be larger than those of accidents at other facilities. Moreover, however rarely accidental releases occur, they may dominate the average contribution of the power plants themselves to human exposures.

---

[1] This plant, owned by Allied-General Nuclear Services, is not available for operation. Although the separations facilities are complete, the plant cannot be licensed until the facilities for conversion of plutonium and waste are completed and unless fuel reprocessing is authorized.

Bibliography — Chapter Three

"NTIS" indicates report is available from: National Technical Information Service, U.S. Department of Commerce, 5285 Port Royal Road, Springfield, VA 22161.

"Basic Radiation Protection Criteria," National Council on Radiation Protection and Measurements report NCRP No. 39 (January 1971).

Discusses background of radiation protection criteria and makes recommendations.

BEIR. "The Effects on Populations of Exposure to Low Levels of Ionizing Radiation," report of the Advisory Committee on the Biological Effects of Ionizing Radiation (the BEIR Committee), National Academy of Sciences — National Research Council (November 1972).

Reviews available evidence on the impacts of radiation on biological organisms in order to estimate the effects of low levels of radiation.

"Coastal Effects of Offshore Energy Systems: An Assessment of Oil and Gas Systems, Deepwater Ports, and Nuclear Power Plants off the Coast of New Jersey and Delaware," U.S. Congress Office of Technology Assessment (November 1976) (U.S. Government Printing Office).

Includes a brief discussion of proposed floating PWR nuclear power plants.

Cohen, B. L., "Hazards from Plutonium Toxicity," *Health Physics,* vol. 32, p. 359 (1977).

Argues, on the basis of anticipated handling of plutonium and the available evidence on its toxicity and pathways, that commercial use of plutonium does not constitute a substantial hazard.

Edsall, J. T., "Toxicity of Plutonium and Some other Actinides," *Bulletin of the Atomic Scientists,* vol. 32, p. 26 (September 1976).

Discusses the hazard constituted by plutonium.

"Environmental Statement Related to Construction of Koshkonong Nuclear Plant," (draft) U.S. NRC report NUREG-0079 (August 1976).

Summarizes environmental impact of proposed Koshkonong PWR nuclear plant.

"Final Environmental Statement, 40 CFR 190: Environmental Radiation Protection Requirements for Normal Operations of Activities in the Uranium Fuel Cycle," 2 vols., U.S. EPA report EPA 520/4-76-016 (November 1976).

States case, including background information, for the standard limiting routine emissions from nuclear power operations.

"Final Environmental Statement. Perry Nuclear Power Plant," U.S. AEC report (April 1974).

Summarizes environmental impact of Perry BWR nuclear plants.

"Final Environmental Statement Relating to Operation of Rancho Seco Nuclear Generating Station," U.S. AEC report (March 1973).

Summarizes environmental impact of the Rancho Seco PWR nuclear plant.

Ford Foundation/MITRE Corporation, "Nuclear Power: Issues and Choices," Report of the Nuclear Energy Policy Study Group (Ballinger, Cambridge, Mass., 1977).

Briefly considers environmental impacts of nuclear plants in the context of an energy policy study.

Gofman, J. W., and Tamplin, A. R., *Poisoned Power* (New American Library, New York, 1974).

Strongly criticizes nuclear power on the basis of its impacts on human health.

Nero, A. V. and Wong, Y. C., "Radiological Health and Related Standards for Nuclear Power Plants," Lawrence Berkeley Laboratory report LBL-5285 (January 1977) (NTIS).

Summary of power plant emissions, of applicable radiation protection standards, and of the current understanding of the effects of radiation on human health.

"Nuclear Power and the Environment," rev. ed. American Nuclear Society, Hinsdale, Ill. (1976).

Environmental impacts of nuclear power, summarized by the society of nuclear power professionals.

NUREG-0116. "Environmental Survey of the Reprocessing and Waste Management Portions of the LWR Fuel Cycle," (supplement 1 to WASH-1248), U.S. NRC report NUREG-0116 (October 1976) (NTIS).

Estimates environmental impacts of fuel reprocessing and waste management for the LWR fuel cycle.

ORNL-4451. "Siting of Fuel Reprocessing Plants and Waste Management Facilities," Oak Ridge National Laboratory report ORNL-4451 (July 1971) (NTIS).

Estimates local and worldwide alteration of radionuclide concentration as a result of fuel reprocessing and waste management operations.

Pigford, T. H., "Environmental Aspects of Nuclear Energy Production," *Annual Review of Nuclear Science,* vol. 24, p. 515 (1974).

Characterizes emissions from commercial nuclear facilities.

Pigford, T. H., et al. "Fuel Cycles for Electric Power Generation," Teknekron report EEED 101 (January 1973, rev. March 1975); "Fuel Cycle for 1000-MW Uranium-Plutonium Fueled Water Reactor," Teknekron report EEED 104 (March 1975); "Fuel Cycle for 1000-MW High-Temperature Gas-Cooled Reactor," Teknekron report EEED 105 (March 1975). These are included in "Comprehensive Standards: The Power Generation Case," U.S. EPA report PB-259-876 (March 1975) (NTIS).

Estimates radionuclide effluents from normal operation of various nuclear systems.

"Review of the Current State of Radiation Protection Philosophy," National Council on Radiation Protection and Measurements report NCRP No. 43 (January 1975).

Reviews philosophy behind setting of radiation protection standards, particularly the question of linear dose-response.

Richmond, C. R., "Current Status of the Plutonium Hot-Particle Problem," *Nuclear Safety,* vol. 17, p. 464 (1976).

Reviews work pertaining to the possibility that plutonium, because it may be present as small particles, may constitute a greater hazard than it would otherwise.

UNSCEAR. "Ionizing Radiation: Levels and Effects," report of the United Nations Scientific Committee on the Effects of Atomic Radiation (the UNSCEAR Committee), United Nations (1972).

Reviews the present levels of radiation and the evidence on the effects of radiation.

WASH-1209. "The Potential Radiological Implications of Nuclear Facilities in a Large Region of the United States in the Year 2000," U.S. AEC report WASH-1209 (January 1973) (NTIS).

Estimates impact of a large nuclear power system.

WASH-1248. "Environmental Survey of the Uranium Fuel Cycle," U.S. AEC report WASH-1248 (April 1974) (NTIS).

Estimates environmental effluents from the LWR fuel cycle.

WASH-1250. "The Safety of Nuclear Power Reactors (Light-Water Cooled) and Related Facilities," U.S. AEC report WASH-1250 (July 1973) (NTIS).

Summarizes radiation protection standards and the impact of nuclear power plants.

WASH-1258. "Final Environmental Statement Concerning Proposed Rule Making Action: Numerical Guides for Design Objectives and Limiting Conditions for Operation to Meet the Criterion 'As Low as Practicable' for Radioactive Material in Light-Water-Cooled Nuclear Power Plant Effluents," 3 vols., U.S. AEC report WASH-1258 (July 1973).

States considerations in setting numerical guides for effluents from nuclear power plants.

WASH-1359. "Plutonium and other Transuranium Elements: Sources, Environmental Distribution, and Biomedical Effects," U.S. AEC report WASH-1359 (December 1974) (NTIS).

Summarizes, in a number of articles, the sources of plutonium, its distribution in the environment, and its biomedical effects.

Wilson, R., and Jones, W. J., *Energy, Ecology, and the Environment* (Academic, New York, 1974).

Briefly discusses radionuclide emissions from reactors and their importance.

CHAPTER FOUR

The Potential for Nuclear Accidents

THE CORE of an operating nuclear power reactor contains a large amount of radioactivity which, if ineffectively contained, can cause substantial harm to humans, either plant workers or surrounding populations. Unusually strict measures are necessary to ensure this containment, since the basic purpose of a reactor is to produce energy and this very energy can be an effective agent for releasing and dispersing radioactivity.

Under normal reactor operating conditions, a balance is achieved between production of heat in the fuel and removal of heat by the coolant. Fuel overheating can occur if the fuel production rate increases enough above normal to exceed the available heat removal capacity or if the capacity of the heat removal system decreases below that required by circumstances. The distinction between these conditions is not unambiguous. Often the first, an unplanned excursion in the reaction rate, is referred to as a "transient." Strictly speaking, though, this word refers to any deviation of operating parameters from their normal values. This includes, therefore, the second class just mentioned, the loss of heat removal capacity. Loss of the coolant from the primary system is the type of reduced removal capacity that is given most attention; such an event is referred to as a "loss of coolant accident" (LOCA).

Some of the basic reactor safety considerations were introduced in Chapter 1. The fundamental design goal is to prevent any transient, including a LOCA, from leading to damage to the fuel, particularly breaching of the fuel cladding or melting of the fuel. Two capabilities are fundamental to preventing such damage: the ability to shut down the chain reaction rapidly and dependably when required, and cooling systems with enough redundancy and capacity to carry away the heat generated in the nuclear core. In certain cases, the basic reactor concept contributes to these emergency capabilities. In others, the basic concept must be augmented by engineered safety features.

Should substantial damage to the core occur, subsidiary systems must exist to control releases of radioactivity to the general environment. The most serious damage would include melting of the core and breach of the reactor vessel. To limit the effects of fuel damage, including core meltdown, the reactor vessel and immediately associated piping are often enclosed in a containment building designed to limit, at least for a time, access of radioactive materials to the atmosphere. Subsystems are often incorporated to control pressure increases within this building and to wash or filter radioactivity from its internal atmosphere.

These remarks apply to almost every type of commercial reactor system. However, most regulatory and public attention has been given to light-water reactors, the current form of commercial power reactor in the United States. The remainder of this chapter will deal explicitly with LWRs and particularly with publicly available assessments of their safety characteristics.

SAFETY DESIGN

The uninterrupted availability of shutdown systems and cooling capacity would ensure that large releases of radioactivity do not occur from a light-water reactor. Limited fuel failures, such as the breach of one or a few fuel rods, are acceptable, since there is no mechanism for such failures to propagate and they cannot, therefore, lead to large-scale failures. It is only when substantial failures occur in the control or cooling systems that the danger of large releases arises. Ultimately, it is failure of cooling that is to be avoided, whether from a failure that originates in the cooling equipment or from a failure (such as an electronic control failure) that leads to a cooling failure.

In the event of a rapid and relatively complete loss of cooling in a reactor that has been operating at design power, large portions of the fuel can quickly lose their structural integrity. The fuel cladding maintains fuel rod structure. When external cooling is removed suddenly, the cladding temperature will rise because of heat initially present in the fuel and because of additional heat that is generated from the roughly 15 billion Ci inventory (of an operating 1000 MWe plant). This radioactive inventory imposes a relatively long-term requirement on the available heat removal capacity. The associated thermal power is shown schematically in Figure 4-1. For the first few minutes after shutdown, the power level is several percent of the value during full power operation. Even at a fraction of a percent level, water must be available to prevent melting.

If large-scale melting of the fuel cladding occurs, significant quantities of fuel can be deposited at the bottom of the reactor core. Unfortunately, it need not stop there. The fuel can reach temperatures so high that it melts its way through the reactor vessel and, eventually (within a day or so), eats its way through the concrete base of the containment building.

Even before fuel escapes from the vessel, a serious state of affairs exists, because breach of the fuel cladding and rise of the fuel temperature release and drive off volatile fission products, most importantly iodine, cesium, and the noble gases, which together comprise a significant portion of the radioactive inventory. Should the containment be breached, particularly by overpressure, large portions of

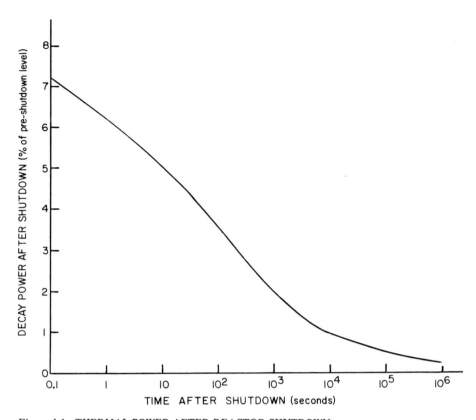

Figure 4-1. THERMAL POWER AFTER REACTOR SHUTDOWN.
After the nuclear chain reaction ceases, radioactivity remaining in the fuel will generate heat as a result of radioactive decay. Assuming that the reactor had been operating for a substantial period, the power generated immediately after shutdown will be approximately 7% of the level before shutdown. For a 3000 MWth reactor, with 1000 MWe capacity, this implies an initial decay power level of about 200 MWth. Due to the rapid decay of short-lived species, this decay heat level decreases rapidly, but is is this heat that imposes the requirement that, in a light-water reactor, cooling water remain available to prevent damage to the fuel.

these volatile radionuclides can escape into the environment. If the breach occurs only by way of the melt-through described above, the earth under the building can provide filtering that is significant, though far from complete.

The basic concept of the reactor, i.e., the fuel design and the control and cooling systems, ensures confinement of the bulk of the radioactivity to the fuel, unless an accident of the type just described occurs. Substantial efforts are expended to prevent failures of the basic reactor systems, particularly of the cooling system. These efforts include not only conservative design of these systems, but continuous monitoring and inspection to ensure their continued serviceability. In addition, the basic systems are designed with redundancy to assure availability. As a second layer of protection, backup systems are available to serve in the event of failure of the basic systems. These include backup reactivity control systems (i.e., neutron poisons) and several emergency core cooling systems.

The adequacy of these ECC systems is highly controversial. The systems are designed to supply water to the primary system in the event that the coolant inventory decreases. For small leaks, there is little doubt that these systems are adequate, assuming that electrical control systems (such as cables) are intact. For large breaks, the sudden release of pressure and the high temperature of the water can result in sudden formation of steam and expulsion of the coolant from the primary system, often referred to as blowdown. Under these conditions, injection of water into the primary system can be difficult. More to the point, the injected water may not reach the hot fuel, so that dangerous rises in temperature may not be arrested.

The ECC systems for PWRs and BWRs are described in Chapters 5 and 6, respectively. (See especially Figures 5-8 and 6-9.) Typically, the reactor has about three ECC systems, which operate in distinct and independent modes. Ordinarily, two of the systems will operate at low pressure, the condition in which most of the coolant has been lost, and one will operate at high pressure (with lower flow rates than at least one of the low-pressure systems). Most of the systems employ relatively independent piping and pumping elements and, for their activation, require a signal indicating low coolant level or abnormal pressure. An important exception is the "accumulator" for the PWR, which consists of a large tank of pressurized water that is kept out of the primary system by a check valve, which opens if the primary system pressure decreases below the accumulator pressure. Such a "passive" system does not require an emergency signal to initiate operation.

These systems are controversial largely because the conditions in a reactor during a LOCA are difficult to predict, so difficult, in fact, that existing computer models are unable to make accurate predictions, either of internal conditions or of the effectiveness of the ECC systems. As a result, two types of models ("codes") have been developed. The "realistic" models attempt to utilize the best current understanding of hydrodynamic behavior to predict what will happen during a LOCA. The results of the codes are therefore the most accurate prediction that can be made. In spite of this, the results are uncertain because of imprecise knowledge of various conditions and concepts that are utilized in the calculation. For this reason, the results of these "realistic" models are not dependable enough, in themselves, for use in the design and licensing of reactors. A second set of "conservative" models has been developed. In areas of uncertainty, these computer codes make assumptions on the "safe" side; i.e., parameters and techniques are chosen that tend to overestimate the severity of conditions, so that the result of the calculation is an accident that is, if anything, more severe than that which would actually occur. Such "conservative" codes form an important part of the licensing process for nuclear power plants. Whether they are really conservative has been highly controversial, as have the small-scale tests designed to check LOCA codes.

ACCIDENT INITIATORS: PIPEBREAKS, EARTHQUAKES, AND PEOPLE

Safety systems are designed to protect the integrity of the reactor system in a wide variety of circumstances. The events that may begin an accident sequence are therefore specifically considered in safety design. Transients can be initiated by many kinds of events, including equipment failures and human failures. Some of these initiators deserve specific mention because of their prominence.

Failure of the components making up a nuclear power plant constitutes one of the main classes of transient initiators. Any device or component has a finite probability of failing during operation, and the plant design has to accommodate these failures while maintaining the integrity of the reactor. In many cases, devices and systems are duplicated or are backed up by other types of systems. In general, the quality of safety-related equipment is required to be unusually high (see next section). However, certain types of failures could really put plant systems to the test. Those given most attention, particularly in pressurized-water reactors, are rupture of pipes, particularly in the primary cooling system, and even of the reactor vessel. Extremely stringent materials, fabrication, and inspection requirements are applied to ensure that the probability of vessel rupture is so low as to contribute little risk. The requirements on piping are also high, but it is expected that pipes will fail on rare occasions. However, auxiliary and safety systems are intended to handle such transients, whereas vessel rupture will necessarily lead to core melting.

External circumstances may also lead to system irregularities. The transient given most attention is an earthquake, although others, such as tornados, broken blades from the steam turbine, and airplane crashes, can stress the reactor system. However, the containment building provides some protection against most external agents. This is not so true of earthquakes, which stress the internal components independently of the existence of an external building. Nuclear power plants are constructed to meet numerous criteria, including the ability to withstand earthquakes. For a given site, the specific criteria depend on a judgment of the expected strength and probability of earthquakes at that site. This is a difficult judgment, requiring identification of the size and activity of nearby faults. Figure 4-2 indicates average seismicity in the contiguous United States. Once the earthquake to be withstood has been determined, the plant is engineered to survive the shock, assuming the postulated earthquake is not excessively large. This requires incorporation of suitable structures and materials into the plant design. The most difficult situation arises when a plant is designed to one standard and it is later decided, on the basis of new geological evidence, that a larger earthquake needs to be considered.[1] To a certain extent, a plant may be refit to new requirements, but there are limits.

A third important category to be mentioned is that of human failures, by which we mean the failure of plant personnel to operate the plant properly, under either ordinary or irregular conditions. Efforts are made to design the plant to withstand most such failures, and personnel training is an important requirement made on the owner of the plant. However, as in other instances, some probability of failure exists. The human actions that would be of most concern are those that could render more than one safety system inoperative. This can bypass the system redundancy that is a basic aspect of safety design, thereby increasing the probability of overall failure.

A final class of some importance is plant maintenance. Certain systems can be

[1] This is the situation for the Diablo Canyon nuclear plants, now being completed in California.

unavailable because of routine testing and maintenance. If one of these is a safety system that is called on during an emergency, it may not be available. Duplication of plant systems reduces the gravity of such circumstances, but this is a possibility that has to be considered in plant design and risk assessment.

SAFETY REVIEW AND QUALITY ASSURANCE

The principal regulator of nuclear power in the United States is the Nuclear Regulatory Commission. The NRC receives this responsibility from the Atomic

Figure 4-2. SEISMIC RISK MAP FOR THE UNITED STATES.
The contiguous United States may be assigned to risk categories as indicated in the figure. These categories correspond to intensities on the Modified Mercalli Intensity Scale, as indicated. (Figure origin: "United States Earthquake, 1968," U.S. Department of Commerce, U.S. Coast and Geodetic Survey, 1970.)

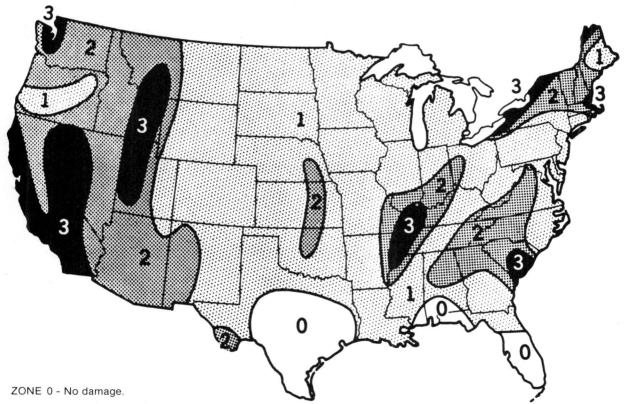

ZONE 0 - No damage.

ZONE 1 - Minor damage: distant earthquakes may cause damage to structures with fundamental periods greater than 1.0 seconds; corresponds to intensities V and VI of the M.M.° Scale.

ZONE 2 - Moderate damage: corresponds to intensity VII of the M.M.° Scale.

ZONE 3 - Major damage: corresponds to intensity VIII and higher of the M.M.° Scale.

TABLE 4-1.

Estimates of Radiological Consequences of Postulated Accidents [a]

NRC Accident Class	Event	Estimated Fraction of 10 CFR Part 20 Limit at Site Boundary[b]	Estimated Dose to Population in 50-mile Radius (man-rem)
1	Trivial incidents	c	c
2	Small releases outside containment	c	c
3	Radwaste system failures		
3.1	Equipment leakage or malfunction	0.033	6.3
3.2	Release of waste gas storage tank contents	0.13	25
3.3	Release of liquid waste storage contents	0.004	0.7
4	Fission products to primary system (BWR)	d	d
5	Fission products to primary and secondary systems (PWR)		
5.1	Fuel cladding defects and steam generator leaks	c	c
5.2	Off-design transients that induce fuel failure above those expected and steam generator leak	<0.001	<0.1
5.3	Steam generator tube rupture	0.043	8.4
6	Refueling accidents		
6.1	Fuel bundle drop	0.007	1.3
6.2	Heavy object drop onto fuel in core	0.12	23
7	Spent fuel handling accident		
7.1	Fuel assembly drop in fuel rack	0.004	0.8
7.2	Heavy object drop onto fuel rack	0.017	3.3
7.3	Fuel cask drop	0.10	20

Energy Act of 1954 (as amended), which originally charged the Atomic Energy Commission with this duty. In addition, state and local agencies are beginning to exercise more authority, although somewhat circumscribed by the Atomic Energy Act, over nuclear power plants.

With the growth of nuclear power to a system in which about 200 commercial nuclear plants have been licensed, or are in the process of being licensed, the regulatory apparatus has grown to be quite substantial. The process is also becoming more and more formal in the sense that the criteria by which licensability is judged are becoming more and more explicit. In another sense, alluded to earlier, the process has an air of formality: the methods by which accident sequences and associated design requirements are calculated are carefully prescribed with the intention that they be conservative, even though they may not be realistic.

The context in which a judgment on the "safety" of a proposed nuclear power plant is made is the review of the Safety Analysis Reports which the applicant must submit before the construction of the plant and just before it is licensed.

TABLE 4-1.

Estimates of Radiological Consequences of Possible Accidents[a]
(continued)

NRC Accident Class	Event	Estimated Fraction of 10 CFR Part 20 Limit at Site Boundary[b]	Estimated Dose to Population in 50-mile Radius (man-rem)
8	Accident initiation events considered in design basis evaluation in the SAR		
8.1	Loss-of-coolant accidents		
	Small break	0.074	25
	Large break	0.56	655
8.1(a)	Break in instrument line from primary system that penetrates the containment	d	d
8.2(a)	Rod ejection accident (PWR)	0.056	65
8.2(b)	Rod drop accident (BWR)	d	d
8.3(a)	Steamline breaks (PWR's outside containment)		
	Small break	<0.001	<0.1
	Large break	<0.001	<0.1
8.3(b)	Steamline break (BWR)	d	d

a. The doses calculated as consequences of the postulated accidents are based on airborne transport of radioactive materials resulting in both a direct and an inhalation dose. Our evaluation of the accident doses assumes that the applicant's environmental monitoring program and appropriate additional monitoring (which could be initiated subsequent to a liquid release incident detected by in-plant monitoring) would detect the presence of radioactivity in the environment in a timely manner such that remedial action could be taken if necessary to limit exposure from other potential pathways to man.

b. Represents the calculated fraction of a whole-body dose of 500 mrem, or the equivalent dose to an organ.

c. These are evaluated as routine releases.

d. Not applicable to PWR.

Source: Adapted from Environmental Statement for Koshkonong PWR.

The NRC prescribes in detail the information and analysis that must be presented in the reports. It also publishes a detailed description (the Standard Review Plan) of the manner in which it reviews these submissions.

The criteria and standards that are used in this review take a number of forms. The fundamental ones, from the point of view of the NRC, are the criteria that are contained in Title 10 of the Code of Federal Regulations (CFR), the title that specifies NRC responsibilities. Part 50 of 10 CFR specifies a number of general design criteria, which apply both to the routine operational characteristics of reactor systems and to their accident response. In many cases, the NRC has formulated more detailed "Regulatory Guides," which specify methods by which the applicant can satisfy these general criteria. Although these Regulatory Guides do not carry legal force, they serve practically as regulations because they turn out to be by far the most convenient way to satisfy the NRC. Finally, the Regulatory Guides, and indeed the entire safety design review, rests substantially on the voluntary standards that various professional societies have formulated for the use of their members and of industry. In many cases, these standards have been formulated for general industrial use; on the other hand, an increasingly large number of standards that are intended for use by the nuclear industry are being published.

The criteria used in the safety review pertain to a number of areas, including materials choice and testing, equipment design, analysis techniques, and quality assurance. The last two areas deserve special comment in this discussion. We have already noted that the accident analysis required for the safety review is carefully specified and somewhat formal. The review requires that a broad range of accidents be considered, from minor disturbances to substantial breakdowns. The accidents are formally designated Class 1 through 8, and a sample listing is included in Table 4-1. For the most serious of these accidents, the LOCA, the required analytical techniques, in the form of the conservative ECC evaluation models discussed above, are specified. The accident analysis must show that emergency systems are capable of dealing with this entire class of accidents without serious damage to the core.

In addition to this detailed analysis of the performance of the reactor's emergency systems, the applicant must calculate the radiation doses that may be received by members of the general public in the event of a reactor accident. This analysis is associated, in particular, with examination of the adequacy of the site proposed for the plant. The models to be used in this calculation are carefully specified, as is the size of the radioactive release to be assumed. The site chosen must be such that an adequate exclusion area (to which the public does not have free access) and low population zone (LPZ) can be specified. The LPZ must have a population small enough to be evacuated if necessary and, furthermore, both the exclusion area and the LPZ must be physically large enough for the maximum radiation doses caused at their boundaries by the assumed radioactive release not to exceed the limits that are given in 10 CFR Part 100.[2] Note that the applicant is not required to calculate realistically the risk that accidents pose for surrounding populations; rather, the utility has to show that the proposed design and site satisfy a number of formal requirements. How the risk, i.e., the probability of accidents times their consequences, has been assessed is discussed in the next section.

Assurance that power plants are adequately safe depends not only on adequate design practice and analytical techniques, but also on assurance that the assumed design is the one that is actually followed. This quality assurance extends to two broad areas: assurance that design and construction of the plant are carried out in a manner that is consistent with practices required for nuclear power plants, and assurance that the components used for the plant meet the quality requirements that were presumed in the design. Quality assurance depends on a structure for inspection and auditing, carried out largely by the utility and its agents, but monitored by the NRC. Specification of the quality assurance program is one of the requirements of the Safety Analysis Report.

Adequacy of the system for regulating nuclear power has been a subject of public debate. The controversy has centered, in the past few years, on the two areas

---

[2] For the LPZ outer boundary, the limit is 25 rem whole body and 300 rem to the thyroid during the entire course of the accident. For the exclusion zone, the dose limit is the same, but for a two-hour time period.

just mentioned: the adequacy of ECC systems and the associated techniques for predicting LOCA behavior and the adequacy of the quality assurance program that makes certain that nuclear plants are designed and built as they are supposed to be. Public pressure has resulted in more severe requirements for the performance of ECC systems. (See the Union of Concerned Scientists report on LOCA criteria and the American Physical Society report on the safety of light-water reactors.) Public attention to quality assurance has heightened since a major fire at the nuclear plants at Brown's Ferry in Alabama, a fire that was caused by sloppy design and construction practice and that could have resulted in fuel melting from loss of coolant. A substantial part of the uncertainty about whether nuclear power plants are sufficiently safe arises from questions about the regulation of nuclear power and, in particular, about the strength of the quality assurance programs associated with nuclear power plants. An indication of many of the safety questions that have arisen in connection with nuclear power plants is given in Table 4-2.

Substantial effort has recently been given to standardizing plant designs and to streamlining the associated licensing procedures. Associated with this effort has been the hope that standardization can reduce the long time required between proposal of a plant and its beginning of operation. Standardization could also reduce the chance of design errors creeping in as changes are made from one plant to the next and could reduce fabrication failures by incorporating identical components and systems in many plants. Whether standardization will succeed in any of these goals remains to be seen: there are suggestions that it will even lengthen the licensing process, unlikely as that may appear.

RISK ASSESSMENT

It is clear that accidents at nuclear power plants are possible. The important question, however, is not whether they are possible, but how likely accidents are and what their consequences would be. Knowing the likelihood and consequences of possible accidents leads to an understanding of the risk from nuclear power plants. This may then serve as a basis for judging whether such plants are adequately safe.

Before the last decade, the nuclear industry, the utilities that were purchasing nuclear power plants, and, indeed, the agency that both regulated and promoted nuclear power, the Atomic Energy Commission, had a common attitude toward the importance of accidents at nuclear power plants. Based on their understanding of the operation of nuclear power plants, and of the safety systems that they incorporated, these groups regarded the probability of major nuclear accidents to be so small that such accidents posed no serious risk to the public. Although it was clear that, if a reactor core melted and a significant fraction of its radioactivity escaped, serious harm could result, it was presumed that such an event would never occur. The only public AEC report on the risk from reactor accidents (WASH-740) found that a release of half the core's fission products could result in the death of thousands from acute radiation exposure, but experts ventured the opinion that such an accident would occur only once every one hundred thousand to one billion years of

TABLE 4-2.

Rough Categorization of ACRS Generic Items Relating to Light-Water Reactors[a]

ECCS AND LOCA RELATED ITEMS, INCLUDING CONTAINMENT RESPONSE

I-1 Net positive suction head for ECCS pumps

I-3 Hydrogen control after a Loss-of-Coolant Accident (LOCA)

I-20 Capability of biological shield withstanding double-ended pipe break at safe ends

IA-5 ECCS capability of current and older plants

IB-3 Performance of critical components (pumps, cables, etc.) in post-LOCA environment

IB-4 Vacuum relief valves controlling bypass paths on BWR pressure suppression containment

II-2 Effective operation of containment sprays in a LOCA[b]

II-8 BWR recirculation pump overspeed during LOCA

II-10 Emergency Core Cooling System capability for future plants

IIA-1 Pressure in containment following LOCA[b]

IIA-3 Ice condenser containments

IIA-5 PWR pump overspeed during a LOCA

IIB-3 Behavior BWR Mark III containments[b]

IIC-1 Locking out of ECCS power-operated valves

IIC-5 Vessel support structures

IIC-8 Behavior of BWR Mark I containments

QUALITY ASSURANCE, INSPECTION, TEST, AND MONITORING

I-9 Vibration monitoring of reactor internals and primary system

I-11 Quality assurance during design, construction, and operation

I-12 Inspection of BWR steam lines beyond isolation valves

I-15 Pressure vessel surveillance of fluence and shift

I-18 Criteria for preoperational testing

I-23 Quality group classifications for pressure retaining components

I-25 Instrumentation to detect stresses in containment walls

IA-2 Primary system detection and location of leaks

IB-2 Fixed in-core detectors on high-power PWRs

II-4 Instruments to detect fuel failures[b]

II-5 Monitoring for excessive vibration or loose parts inside the pressure vessel

II-11 Instrumentation to follow the course of an accident

IIA-8 ACRS/NRC periodic 10-year review of all power reactors

IIC-7 Maintenance and inspection of plants

EMERGENCY CONTROL

I-2 Emergency power

IA-4 Anticipated transients without scram

IB-5 Emergency power for two or more reactors at the same site

IB-7 Control rod ejection accident

IIA-2 Control rod drop accident (BWRs)[b]

PROTECTION AGAINST SABOTAGE

I-8 Protection against industrial sabotage

IIC-3 Design features to control sabotage

GENERAL EQUIPMENT AND SYSTEM ADEQUACY AND PROTECTION

I-6 Fuel storage pool design bases

I-7 Protection of primary system and engineered safety features against pump flywheel missiles

I-13 Independent check of primary system stress analysis

I-14 Operational stability of jet pumps

I-19 Diesel fuel capacity

I-24 Ultimate heat sink

IA-1 Use of furnace sensitized stainless steel

IA-2 Primary system detection and location of leaks

IC-1 Main steam isolation valve leakage of BWR's

IC-2 Fuel densification

II-1 Turbine missiles

II-6 Common mode failures[b]

II-7 Behavior of reactor fuel under abnormal conditions

IIA-4 Rupture of high-pressure lines outside containment[b]

IIA-6 Isolation of low-pressure from high-pressure systems[b]

IIA-7 Steam generator tube leakage

IIB-2 Qualification of new fuel geometries

IIB-4 Stress corrosion cracking in BWR piping

IIC-2 Fire protection

IIC-6 Water hammer

SEISMIC RESPONSE

I-5 Strong motion seismic instrumentation

TABLE 4-2.
Rough Categorization of ACRS Generic Items Relating to Light-Water Reactors[a]
(continued)

I-22	Seismic design of steam lines	I-21	Operating one plant while other(s) is/are under construction
IC-4	Seismic category I requirements for auxiliary systems	IB-1	Positive moderator coefficient
II-9	The advisability of seismic scram[b]	IC-3	Rod sequence control systems
		IIB-1	Hybrid reactor protection system
REACTOR PRESSURE VESSEL		**EFFLUENTS AND DECONTAMINATION**	
I-10	In-service inspection of reactor coolant pressure boundary	IB-6	Effluents from light-water-cooled-nuclear power reactors
I-16	Nil ductility properties of pressure vessel materials	IIC-4	Decontamination and decommissioning of reactors
II-3	Possible failure of pressure vessel post-LOCA by thermal shock		
GENERAL REACTOR OPERATION: CONTROL AND INSTRUMENTATION			
I-4	Instrument lines penetrating containment		
I-17	Operation of reactor with less than all loops in service		

a. Class I items are "resolved"; class II are not. A, B, and C indicates, respectively, items that were added in the second, third, and fourth ACRS reports.
b. Items considered resolved by the NRC staff but pending by the ACRS.
Source: A. V. Nero and M. R. K. Farnaam, "A Review of Light-Water Reactor Safety Studies," Lawrence Berkeley Laboratory report LBL-5286 (January 1977). These are the items that had been brought up for consideration by mid-1976. The Advisory Committee on Reactor Safeguards (ACRS) is a panel that advises the Nuclear Regulatory Commission on matters of licensing and reactor safety.

reactor operation. It has not been uncommon to estimate that core meltdowns would have a rate of occurrence of once per million reactor years (i.e., cumulative years of reactor operation).

The empirical evidence says nothing about the probability other than that a meltdown has not occurred in the about 300 reactor years of commercial reactor operation that this country has so far experienced. Inclusion of power reactors on naval vessels would raise this to about 2000 reactor years, presuming that experience with such reactors is considered to be pertinent. This obviously says little about events that are thought to occur extremely rarely. On the other hand, many critics of nuclear power regard the Brown's Ferry incident to have been a "near meltdown" accident.

Only during the 1970s have serious attempts been made to predict on mechanistic grounds the probability of reactor accidents. The prototype of such calculations is the "Reactor Safety Study" performed for the NRC by a group under the direction of Professor Norman Rasmussen (of Massachusetts Institute of Technology). The results of this group's extensive work were published in 1975 in the form of an NRC (AEC) report, WASH-1400; a draft report was made available for public review in 1974. Many aspects of this work have been heavily criticized, but, however faulty, it stands as the most comprehensive effort to calculate the risk from nuclear plant accidents.

Figure 4-3. EVENT TREE.
Accident sequences may be identified using an event tree, as indicated in the figure. Each sequence begins with a specific initiating event, and a branch occurs wherever a safety-related function may or may not operate successfully. If the probability of failure (represented by "P" in the figure) is known for each function, the overall probability of failure may be calculated. (The function failure probability is often calculated using a "fault tree.") In the event tree, certain of the functions are not independent, so that the tree may be reduced to a simpler form. For example, failure of electric power ensures overall failure. (Figure reproduced from WASH-1400.)

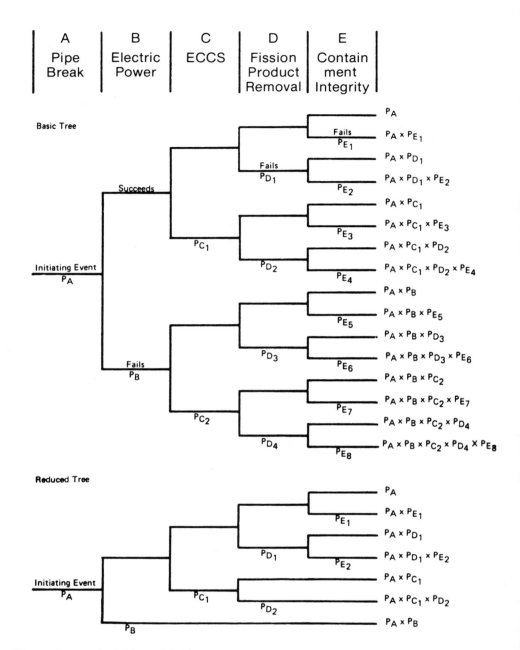

Note—Since the probability of failure, P, is generally less than 0.1, the probability of success (1-P) is always close to 1. Thus, the probability associated with the upper (success) branches in the tree is assumed to be 1.

Such a calculation requires three steps: identification of possible accident sequences, i.e., sequences of events that could lead to a release of radioactivity; calculation of the probability of each accident; and calculation of the consequences of each accident type. The first two steps are largely a characteristic of the nuclear power plant being considered. The consequences, on the other hand, depend not only on the nature of the radioactive release, but also on the location of the plant with respect to nearby population and property and on the meteorological conditions during the accident. (The site characteristics may also affect, to some extent, the probability of specific types of accidents. An important example of this is the effect that a major earthquake could have on the plant.) Once the probability and consequences are known, their importance may be considered in relation to the benefit to be derived from utilization of nuclear power. Alternatively, efforts could then be made to reduce the probability of those accident types that contribute the most risk.

The overall risk may be obtained by summing the probability times consequences for all the contributing accident types. This quantity, given in terms of consequences per year of reactor operation, is a measure of the liability that is accepted in the use of nuclear power. A similar quantity could be estimated for other types of power production and for other activities. The consequences may be of various types, including acute sickness or death due to heavy doses of radiation, latent sickness and death, and property damage.

In the NRC Reactor Safety Study, accidents were identified and their probabilities calculated using moderately new techniques involving what are called "event trees" and "fault trees." As indicated in Figure 4-3, the event tree is an analytical technique that begins with an initiating event, such as a pipebreak, one of the major causes of a LOCA. Presuming such an event occurs, the analyst identifies the various systems and functions (such as electrical service or the emergency core cooling systems) that may prevent damage to the fuel or, failing that, which may minimize the severity of the resulting radioactive release. Beginning with the initiating event, each such system (or function) provides a branch where either success or failure is possible. The tree thus constructed may be used to identify the sequences that lead to failure, i.e., to meltdown and the resulting radioactive release, and also as a guide to characterizing the release, in regard to both time sequence and magnitude. (Actual calculation of the time sequence and magnitude of release requires modeling of the physical processes of importance during the accident.) Similar trees are often used in various kinds of decision-making.

Once the event tree is constructed, the probability of an accident sequence may be calculated if the probability of failure of the various safety systems and functions is known and if the probability of the initiating event is known. It is calculation of the system failure probabilities that often requires use of "fault trees," a tool that had previously been used in reliability analysis for complex systems, such as those used in the space program. Parts of the event tree for failure of electrical service in a PWR are given in Figure 4-4. Such a tree is constructed by asking what could cause the failure of the system of interest. Once the subsystems that could cause failure, because of their failure either independently or in combi-

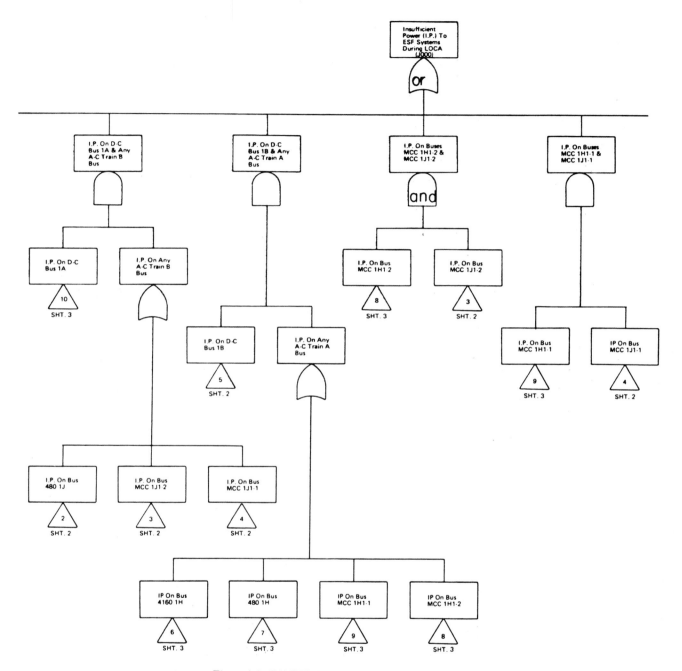

Figure 4-4. SAMPLE FAULT TREE.
The figure shows portions of the fault tree associated with insufficient electrical power to the engineeered safety features of a PWR during a LOCA. The portion of the figure above shows some of the main branches leading to insufficient power; the symbol connecting the branches to the upper box is the symbol for "or," i.e., a failed condition for any one of the branches yields failure of electrical power. The other branch symbol employed is an "and," i.e. failure occurs when all the subsidiary branches have a failed condition. The portions of the figure at the right are portions of the fault tree that are to be connected to the portion above as indicated by the numbered triangles. The entire fault tree for insufficient power is much more extensive than indicated in the figure. (Figure reproduced from WASH-1400.)

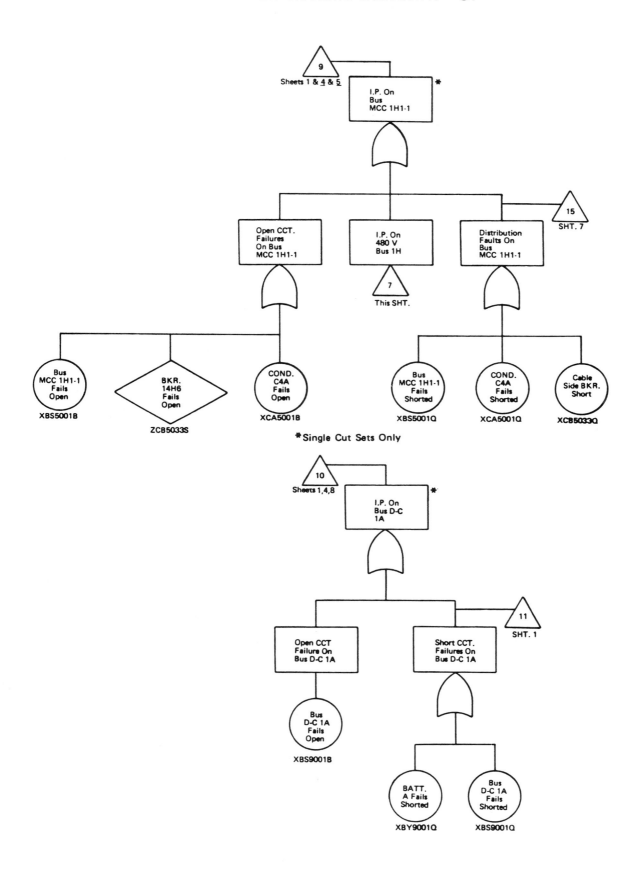

*Single Cut Sets Only

nation, are identified, the analyst asks what other subsidiary systems could cause the failure of these subsystems. This process continues, forming a "tree," until systems, components, or human operators are reached whose probability of failure can be estimated from experience. Assignment of these probabilities from the limited data that are available is highly uncertain and one of the controversial aspects of the Reactor Safety Study. Supplying these probabilities to the fault tree, the system failure probability is obtained for use in calculating the accident probability from the previously constructed event tree. The result is an accident probability that is associated with each accident sequence; and, in addition, each of these sequences is characterized in terms of the timing, magnitude, and type of release.

The next step is to calculate the consequences of each of the accident sequences. For convenience, the Reactor Safety Study did not perform such a calculation for each accident sequence, but rather assigned these sequences to a small number of groups, within each of which the character of the radioactive release was practically the same. For each of these "release categories," the study then calculated a spectrum of consequences that depended on the disposition of populations in relation to the power plant and on the meteorological conditions during the accident. The data on populations and meterological conditions were taken from the sites at which the first 100 commercial nuclear plants are to operate. The spread of radioactivity from the release point was calculated using a standard type of dispersion model. The consequences to the health of the exposed population were calculated using available information on both acute and long-term health effects.

Each step of the analysis performed in the Reactor Safety Study involves uncertainties, and in many cases these uncertainties are large. The uncertainties in the final results are stated to be about a factor of 5 in both the probabilities and the consequences. Many critics believe that the uncertainties are much larger than this.

The study was based on examination of two reactors, one a PWR and one a BWR. The overall results were slightly, not substantially, different for the two types, although the accident sequences would be substantially different for the two types, because of their distinct designs. In the following discussion, no distinction is made between the types. It is interesting to note, though, that much of the criticism of ECC systems has centered on the PWR. The WASH-1400 results avoid this emphasis, perhaps because the study presumed the design adequacy of the ECC systems; i.e., if the ECC systems operated, they would succeed in cooling the core.

The report of the Reactor Safety Study, WASH-1400, displays its results in the form of graphs which plot frequency of accidents (essentially probability per year for 100 reactors) of some minimum consequences versus the consequences. This is done for early fatalities, early illness, latent cancer fatalities, thyroid nodules (an operable condition), genetic effects, and property damage. Figure 4-5 displays the results for early fatalities. Figure 4-6 does the same for latent cancer fatalities. The reader should note, though, that the two figures cannot be compared directly, since Figure 4-5 shows the net consequences from a given accident, while Figure 4-6 shows the consequences *per year* at some time after the accident. To compare the two results, the latent cancer effects must be multiplied by about a factor of 30, the approximate period during which the deaths would occur.

TABLE 4-3.

Wash-1400 Annual Risk Estimates for 100 Nuclear Power Plants

APPROXIMATE AVERAGE SOCIETAL AND INDIVIDUAL RISK PROBABILITIES PER YEAR FROM POTENTIAL NUCLEAR PLANT ACCIDENTS[a]		
Consequence	*Societal*	*Individual*
Early fatalities[b]	3×10^{-3}	2×10^{-10}
Early illness[b]	2×10^{-1}	1×10^{-8}
Latent cancer fatalities[c]	7×10^{-2}/yr*	3×10^{-10}/yr
Thyroid nodules[c]	7×10^{-1}/yr*	3×10^{-9}/yr
Genetic effects[d]	1×10^{-2}/yr	7×10^{-11}/yr
Property damage ($)	2×10^{6}	--

a. Based on 100 reactors at 68 current sites.

b. The individual risk value is based on the 15 million people living in the general vicinity of the first 100 nuclear power plants.

c. This value is the rate of occurrence per year for about a 30-year period following a potential accident. The individual rate is based on the total U.S. population.

d. This value is the rate of occurrence per year for the first generation born after a potential accident; subsequent generations would experience effects at a lower rate. The individual rate is based on the total U.S. population.

* To obtain the net risk of these delayed effects, for comparison with early effects, these entries should be multiplied by 30 (see note c). This yields an annual risk of 2 cancer fatalities and 20 thyroid nodules.

This clarification is important because a failing of WASH-1400 is that it tends to obscure the fact that the estimated latent fatalities are far more numerous than the early fatalities. This is clear from Table 4-3, also taken from WASH-1400, which gives the overall risk (the sum of probability times consequences from 100 operating plants). The average societal risk from 100 plants is 0.003 per year from early fatalities, but for latent fatalities it is 0.07 per year for each of the 30 years of incidence after accidents. Therefore, the number to be compared with 0.003 is actually 30×0.07 or about 2: the overall risk from latent fatalities is thus found to be about 700 times that from early fatalities. Ironically, one would never surmise from the executive summary of WASH-1400 that the risk from latent deaths is even as much as that from early deaths. Not surprisingly, such presentation of the results of the study has been heavily criticized.

A lamentable aspect of the presentation, too, is the extent to which the details of the calculations have been obscured. The study constructed a computer code which internally performed the calculation for various accident categories and for the individual site characteristics and then lumped the results into the form given in Figures 4-5 and 4-6. Attempting to see how the results are affected by alteration of various parameters or how they vary from site to site generally means performing much of the calculation again. Moreover, the discussion in WASH-1400 leaves the reader with some impressions that appear to be inconsistent with the data presented in the report. For example, the report leads one to believe that most of the risk arises from accidents that do not involve large radioactive releases, but that are much more probable than the large releases. On the contrary, a critical examination of the data indicates that, however improbable the very large releases are, the

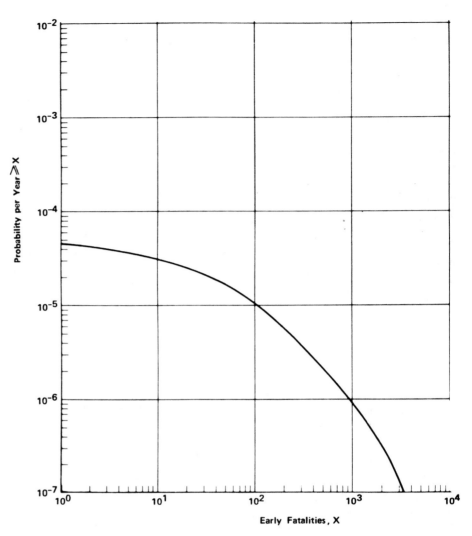

Note: Approximate uncertainties are estimated to be represented by factors of 1/4 and 4 on consequence magnitudes and by factors of 1/5 and 5 on probabilities.

Figure 4-5. EARLY FATALITY ESTIMATES (WASH-1400).
The figure shows the probability that in a given year nuclear plant accidents will cause a minimum of the specified number of early fatalities, i.e., deaths from acute radiation exposure. The probability given assumes that 100 plants, of the type studied by WASH-1400, are operating. Note for example that the probability of one death occurring in a given year was found to be less than one in ten thousand (for 100 reactors). (Figure reproduced from WASH-1400.)

amount of radioactivity escaping is so large that they appear to contribute most of the risk from plant accidents. Likewise, although the report seems to suggest that most of the risk arises from accidents that have small consequences, this too appears to be incorrect. In any case, the manner of presentation of the report does not lend the results to easy interpretation or application.

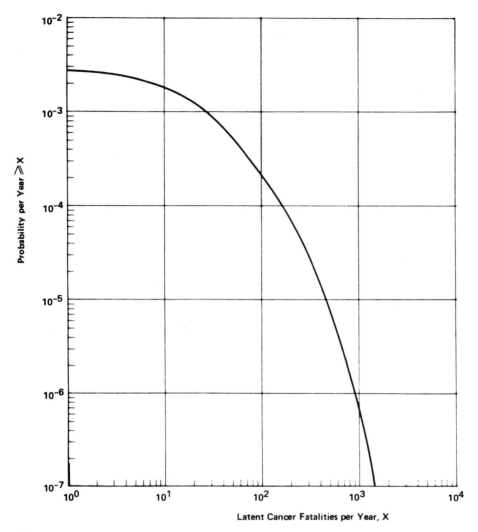

Note: Approximate uncertainties are estimated to be represented by
         factors of 1/6 and 3 on consequence magnitudes and by factors
         of 1/5 and 5 on probabilities.

Figure 4-6. LATENT CANCER FATALITY ESTIMATES (WASH-1400).
The figure shows the probability that in a given year nuclear plant accidents at any of 100
reactors will cause a minimum of a certain number of cancer fatalities *per year* during a
30 year latency period after the accidents. To get the total number of cancer fatalities caused
versus the annual probability of this minimum, the scale at the bottom should be multiplied
by 30. (Figure reproduced from WASH-1400.)

The Reactor Safety Study has been criticized for more than its presentation.
Reviewers have expressed skepticism that all the important accident sequences have
been identified; omissions that could be particularly important would be "common
mode" failures, where one failure renders more than one supposedly independent
safety system inoperative, thereby circumventing the protection that would be

offered by multiple systems. Critics have belittled use of "fault trees" for analysis of low probability events in complex systems. They often express lack of confidence in the human and component failure rate data that was used in calculating the system failure probability; reviewers are often skeptical, too, of the quoted uncertainties, particularly for the final results given by WASH-1400. Finally, substantial criticism has been offered of the manner in which the consequences were calculated, in regard to modeling of both the dispersion of released radioactivity and the health effects of resulting doses.

The Nuclear Regulatory Commission has endeavored to answer many of these criticisms. At the present time, it is still difficult to judge how accurate the Reactor Safety Study results are. Nonetheless, the study does constitute a substantial beginning in nuclear risk assessment. And, however dependable (or not) the results of the Reactor Safety Study may be, they serve as a quantitative guide in examining the risk from nuclear power plants. Overall, the results indicate that, for example, less than one accident death (mostly from cancer) would result from the average year of reactor operation. Even a very critical view of nuclear safety cannot raise this above 10 deaths per reactor year. This would be a significant number. It does not, however, appear to be more serious than the harm accepted from oil- or coal-burning power plants, as they are now operated. It is also useful to note, as was emphasized by a recent Ford Foundation report on nuclear power, that most of the risk calculated in WASH-1400 arose from only two or three plants sited close to large population centers. More distant siting substantially reduces risk; the price paid is greater energy loss during transmission of electricity.

One aspect of reactor safety that such an all-encompassing view obscures is the importance of large accidents versus small ones. As noted, the data of WASH-1400 indicate that most of the risk of death arises from relatively large accidents: most of the risk of early death arises from accidents that would cause hundreds of such deaths, and most of the risk of cancer death arises from accidents that would cause thousands of such deaths during the years after the accident. The largest accidents are also the rarest, but their consequences are so great that they cannot be considered in exactly the same manner as small accidents. Overall risk estimates do not supply the kind of information needed for emergency planning around nuclear power plants; such planning must consider the balance between large and small accidents.

Bibliography — Chapter Four

"NTIS" indicates report is available from: National Technical Information Service, U.S. Department of Commerce, 5285 Port Royal Road, Springfield, VA 22161.

APS-1975. H. W. Lewis et al., "Report to the American Physical Society by the Study Group on Light-Water Reactor Safety," *Reviews of Modern Physics,* vol. 47, supplement no. 1, p. S1 (1975).

Reviews several aspects of reactor safety and the reactor safety research program, emphasizing loss-of-coolant accidents and accident consequences.

Bulletin of the Atomic Scientists, articles on "Nuclear Reactor Safety" by J. Primack, F. C. Finlayson, N. C. Rasmussen, R. K. Weatherwax, H. J. C. Kouts, F. von Hippel, and H. A. Bethe. *Bulletin of the Atomic Scientists,* vol. 31, p. 15 (September 1975).

Several articles, with varying opinions, on nuclear reactor safety.

"Coastal Effects of Offshore Energy Systems: An Assessment of Oil and Gas Systems, Deepwater Ports, and Nuclear Power Plants off the Coast of New Jersey and Delaware," U.S. Congress Office of Technology Assessment, U.S. Government Printing Office (November 1976).

Describes proposed floating PWR nuclear power plants.

Cottrell, W. B., "The ECCS Rule-Making Hearing," *Nuclear Safety,* vol. 15, p. 30 (1974).

Describes the hearings held on criteria for LWR emergency core cooling systems.

"Environmental Statement Related to Construction of Koshkonong Nuclear Plant," (draft), U.S. NRC report NUREG-0079 (August 1976).

Summarizes environmental impact of Koshkonong PWR nuclear plant.

EPRI 248-1. "Heat Transfer During the Reflooding Phase of the LOCA — State of the Art," Electric Power Research Institute report EPRI 248-1 (September 1975) (NTIS).

Describes PWR LOCAs, related safety systems, and computer codes for calculating conditions during accidents.

Ford Foundation/MITRE Corporation, "Nuclear Power: Issues and Choices," Report of the Nuclear Energy Policy Study Group (Ballinger, Cambridge, Mass., 1977).

Briefly reviews the significance of reactor accidents in making choices on nuclear policy.

"General Description of a Boiling Water Reactor," General Electric Company report (March 1976).

Describes safety systems of a BWR power plant.

Kendall, H. W. et al., "The Risks of Nuclear Power Reactors: A Review of the NRC Reactor Safety Study," Union of Concerned Scientists, Cambridge, Mass. (August 1977).

A critique of WASH-1400 (the final version). The Sierra Club and UCS published a critique of draft WASH-1400 in November 1974.

McPherson, G. D., "Results of the First Three Nonnuclear Tests in the LOFT Facility," *Nuclear Safety,* vol. 18, p. 306 (1977).

Summarizes the results of the initial testing program at the loss-of-fluid test facility; a nuclear core had not yet been used.

Nero, A. V. and Farnaam, M. R. K., "A Review of Light-Water Reactor Safety Studies," Lawrence Berkeley Laboratory report LBL-5286, (January 1977) (NTIS).

Summarizes the NRC Reactor Safety Study, the American Physical Society study of LWR safety, and others, including comments on these.

Nuclear Regulatory Commission, "1976 Annual Report," U.S. Government Printing Office (May 1977).

Summary of NRC programs for 1976.

Primack, J. and von Hippel, F., "Nuclear Reactor Safety," *Bulletin of the Atomic Scientists,* vol. 30, p. 5 (October 1974).

Brief summary of the status of the debate over nuclear reactor safety as of a few years ago.

"Reactor Safety Study (WASH-1400): A Review of the Final Report," U.S. EPA report EPA-520/3-76-009 (June 1976).

EPA comments on WASH-1400.

"Systems Summary of a Westinghouse Pressurized Water Reactor Power Plant," Westinghouse Electric Corporation report (1971).

Describes safety systems of a PWR power plant.

WASH-740. "Theoretical Possibilities and Consequences of Major Accidents in Large Nuclear Power Plants," U.S. AEC report WASH-740 (1957) (NTIS).

Calculates the early deaths and property damage that may occur for several types of radioactive releases; does not calculate probabilities.

WASH-1250. "The Safety of Nuclear Power Reactors (Light-Water Cooled) and Related Facilities," U.S. AEC report WASH-1250 (July 1973) (NTIS).

Describes light-water reactor systems and discusses many safety-related questions.

WASH-1400. "Reactor Safety Study: An Assessment of Accident Risks in U.S. Commercial Nuclear Power Plants," 9 vols., U.S. NRC report WASH-1400, NUREG-75/014 (October 1975) (NTIS).

Report of the AEC/NRC-sponsored study on the risks from accidents at light-water reactors; a draft was released for comment in August 1974.

Wessman, G. L., and Moffette, T. R., "Safety Design Bases of the HTGR," *Nuclear Safety,* vol. 14, p. 618 (1973).

Presents major safety features of HTGRs. See also chapter 8 bibliography.

Ybarrondo, L. J., Solbrig, C. W., and Isbin, H. S., "The 'Calculated' Loss-of-Coolant Accident: A Review," American Institute of Chemical Engineers Monograph Series, no. 7, (1972).

Includes a brief description of LWR ECC systems.

Part II

Commercial Nuclear Reactors

A useful distinction may be made between "commercial" and "advanced" nuclear reactors. The first category includes those reactors whose development has progressed to the point where they are available in the marketplace. These reactors, for which utilities have placed or can place orders, constitute the current and near-term basis for commercial nuclear power. On the other hand, "advanced" reactors, those that incorporate significant improvements over commercial designs, will be the basic components in any nuclear power system for the long-term future. A number of advanced reactors which have reached the conceptual design or engineering stages are described in Parts III and IV.

The purpose of this part of the book is to acquaint the reader with the basic features of reactors that have been offered commercially in the United States and Canada. Only domestic firms have sold reactors in these two countries. Since these same companies also hold a significant portion of the international market, a description of their reactors provides a fair introduction to the characteristics of commercial reactors throughout the world. Moreover, even though several countries — notably France, Germany, England, Japan, Sweden and the USSR — have substantial domestic nuclear programs, and even sell their reactors to other countries, most of these are light-water reactors similar to U.S. designs.

The reactors referred to briefly in Chapter 2 — water- or gas-cooled thermal reactors — are described fully in Part II. The discussion covers light-water reactors, both the pressurized-water (PWR) and boiling-water (BWR) varieties, heavy-water reactors, particularly the Canadian (CANDU) reactor, and gas-cooled carbon-moderated reactors, as exemplified by the no longer offered high-temperature gas-cooled reactor (HTGR). In their present forms, these reactors may be called "converters," as an indication of the fact that they convert some fertile material to fissile, but not enough to approach breakeven or produce a profit. (The possibility that modified versions of such reactors could achieve a conversion ratio close to 1 is discussed in Chapters 10 and 14.) The fissile material thus produced, primarily ^{239}Pu or ^{233}U, can then contribute to the chain reaction in the same or (assuming fuel reprocessing) another reactor.

Scores of nuclear power plants are now in operation in the United States and abroad, supplying about 50 gigawatts capacity in the United States and a similar amount elsewhere. As was suggested in Figure 2-3, many more are under construction or in planning. These are almost all water-cooled reactors. In the United States, all but one of the reactors operating or under construction are LWRs. Of these, 60% of those operating and 70% of those under construction are pressurized-water reactors; the remainder of the market is supplied by boiling-water reactors.

CHAPTER FIVE

Pressurized-Water Reactors

MOST OF THE light-water reactor power plants now operating or under construction use pressurized-water reactors. Westinghouse supplies somewhat more than half of the PWRs in the United States, with the remainder split between Babcock & Wilcox and Combustion Engineering. Many of the details of PWRs vary from one vendor to another and even, for the same manufacturer, from one reactor to the next. However, the fundamental characteristic of PWRs remains the same: that the primary coolant raises steam in a heat exchanger called a steam generator and this steam drives the turbine. A basic PWR system is shown schematically in Figure 5-1. Enclosed in a containment structure is the primary coolant system consisting of the reactor vessel and two or more primary coolant loops, each including piping, pumps, and a steam generator (perhaps shared). The safety injection (ECC) systems are also within the containment. Steam from the steam generators is transported out of the containment to the turbo-generator system. Condensate returns to the steam generators. Although three corporations offer PWRs, the system description that follows is based largely on that of Westinghouse. PWRs from the other manufacturers will vary in detail, particularly in the matter of the primary coolant loop arrangement.

BASIC PWR SYSTEM

The basic unit of a PWR core is a fuel pin typical of water-cooled reactors. For such reactors, the uranium dioxide fuel material is pressed into "pellets," cylinders about one-half inch in diameter and of similar height. These pellets are sintered (heated to high temperatures), ground to the proper dimensions, then sealed, along with a helium atmosphere, in a cladding material. This constitutes a fuel rod or pin. The cladding is typically an alloy of zirconium, chosen for its low neutron cross-section, as well as for its structural properties. The fuel pin for a light-water reactor is shown schematically in Figure 5-2. These pins, each more than

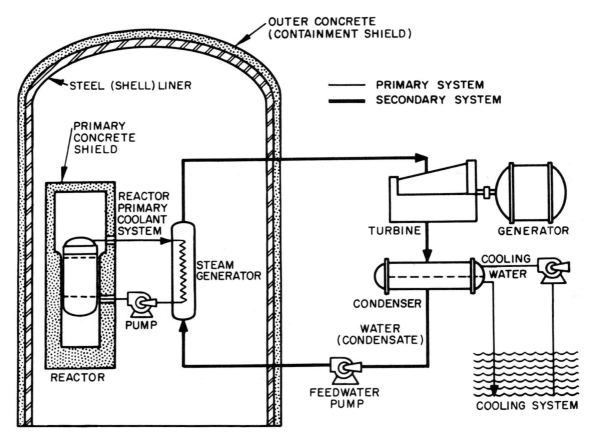

Figure 5-1. SCHEMATIC PRESSURIZED-WATER REACTOR POWER PLANT.
The primary reactor system is enclosed in a steel-lined concrete containment building. Steam generated within the building flows to the turbine-generator system (outside the building), after which it is condensed and returned to the steam generators. (Figure reproduced from ERDA-1541.)

12 feet (3.6 m) long for LWRs, are assembled into bundles or "assemblies," the operational unit for handling, refueling, etc. Should plutonium be recycled into light-water reactors, it would be handled in much the same way. In the United States, it has been proposed that the plutonium oxide be finely mixed with the uranium dioxide before a fuel pellet is formed.

The core of a pressurized water reactor consists of a large number of square fuel assemblies or bundles. Figure 5-3 shows one of these assemblies, in this case containing a control rod cluster. Many PWRs use assemblies that consist of 15 × 15 arrays of fuel pins of the type indicated in Figure 5-2, each somewhat more than 12 feet long. Newer PWRs use 17 × 17 assemblies. These pins or rods are closely held together in a matrix with no outer sheath, by the assembly's top and bottom structures, and by spring clip grid assemblies. A full-sized (about 1000 MWe) PWR may contain nearly 200 assemblies with about 40 or 50 thousand fuel pins, contain-

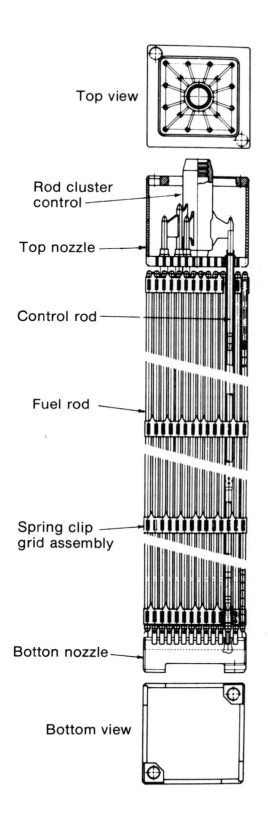

Top view

Rod cluster control

Top nozzle

Control rod

Fuel rod

Spring clip grid assembly

Botton nozzle

Bottom view

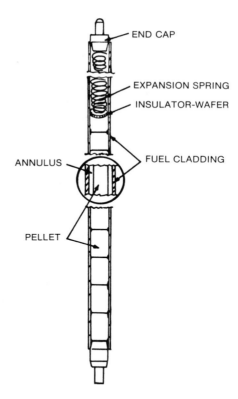

END CAP

EXPANSION SPRING

INSULATOR-WAFER

ANNULUS

FUEL CLADDING

PELLET

Figure 5-2. CUTAWAY VIEW OF OXIDE FUEL FOR COMMERCIAL LWR POWER PLANTS.

The basic unit in the core of a light-water reactor is a fuel rod containing uranium oxide pellets in a Zircaloy cladding. The rod is filled will helium gas and welded shut. The circled portion exaggerates the annular space between the pellet and the cladding. (Figure reproduced from WASH-1250.)

Figure 5-3. FUEL ASSEMBLY FOR A PRESSURIZED-WATER REACTOR.

In a pressurized-water reactor, fuel rods are assembled into a square array, held together by spring clip assemblies and by nozzles at the top and bottom. The structure is open, permitting flow of coolant both vertically and horizontally. All the assemblies in the reactor may have the same mechanical design, including provision for passage of a control rod cluster (shown in the figure). Where there is no cluster, these positions may have neutron sources, burnable poison rods, or plugs. (Figure reproduced from WASH-1250.)

ing about 110 tons (100 metric tons) of uranium dioxide (and plutonium, were recycle to occur).

All the assemblies have provision for the passage of control rods through rod guides which take about 20 of the positions that could otherwise hold fuel rods. If the assembly is used as a control assembly, and about 30% of them are, the rods from that assembly are manipulated from the top as a cluster. The control drives are at the top of the pressure vessel. In case the assembly does not contain a rod cluster, control rod positions may be taken by burnable poison, in this case boron 10 which is used after initial reactor operation to offset excess reactivity, or by neutron sources, used for reactor startup. Otherwise, these positions are left vacant and water flow through them is blocked.

Most of the control rods have silver-indium-cadmium neutron absorber for the full length of the core and are used for operational control of the reactor, including load following, and for quick shutdown capability. Reactor "trip" capability is provided by the fact that the rods can simply be dropped into place gravitationally; somewhat fewer than half the control assemblies are reserved for this shutdown capability, the remaining being used for operational control. Some of the control rods have absorber only in their bottom quarter and are used for shaping the axial (vertical) power distribution. The other basic means of control is to introduce boric acid into the primary coolant. This method is used both for shutdown and for adjusting the reactivity to take account of long-term changes, such as reduction in fissile content and buildup of fission product poisons. Effectively, boron adjustment is used to keep the reactivity within the range of the control rods.

The core has three enrichment zones, with the most highly enriched (slightly greater than 3%) at the periphery and the other enrichments scattered through the interior, all to provide a relatively flat power distribution. The average power generation density in the core is about 98 kW/liter. (See Table 5-1 for other PWR parameters.) This energy is carried away by a very large flow of water, about 140 million pounds per hour (18 Mg/s). The water's operating temperature is about 600 °F (315 °C), which maintains the clad temperature nominally below 700 °F (371 °C).

The core, control rods, and core-monitoring instrumentation are contained in a large pressure vessel, designed to withstand pressures, at operating temperatures, of about 2500 psi (17 MPa). The vessel may be about 40 feet in height (12 m) and 14 feet (4 m) in diameter, with carbon steel walls 8 inches (20 cm) or more thick. All inner surfaces that come into contact with the coolant are clad in stainless steel. (This is also true of all other parts of the primary coolant system, except for those portions that are made of Zircaloy or Inconel, i.e., the fuel cladding and the steam generator tubing, respectively.) The top head of the vessel, which holds all the control rod drives, is removable for refueling. The reactor vessel and its contents are shown in Figure 5-4.

The coolant enters the reactor vessel through nozzles near the top of the core and, constrained by a "core barrel" between the vessel and the core, flows to the bottom of the core. The water then flows up through the core and out exit nozzles

TABLE 5-1.
Representative Characteristics of Pressurized-Water Reactors

Core thermal power	3,411 MWth
Plant efficiency	32%
Plant electrical output	1,100 MWe
Core diameter	134 in (3.4 m)
Core (or fuel rod) active length	144 in (3.7 m)
Core weight (mass)	276,000 lb (125 Mg)
Core power density	98 kW/liter
Cladding material	Zircaloy-4
Cladding diameter (OD)	0.422 in (1.07 cm)
Cladding thickness	0.024 in (0.06 cm)
Fuel material	UO_2
Pellet diameter	0.37 in (0.9 cm)
Pellet height	0.6 in (1.5 cm)
Assembly array	15 × 15, open structure[a]
Number of assemblies	193
Total number of fuel rods	39,372[a]
Control rod type	B_4C or Ag-In-Cd in cylindrical rod
Number of control rod assemblies	60 (may vary considerably)
Number of control rods per control assembly	20 (may vary considerably)
Total amount of fuel (UO_2)	217,000 lb (98 Mg)
Fuel power density	38 MW/Te
Fuel/coolant ratio	1/4.1
Coolant	Water (liquid phase)
Total coolant flow rate	136 × 10⁶ lb/hr (17 Mg/sec)
Core coolant velocity	15.5 ft/sec (4.7 m/sec)
Coolant pressure	2,250 psi (15.5 MPa)
Coolant temperature (inlet at full power)	552 °F (289 °C)
Coolant temperature (outlet at full power)	617 °F (325 °C)
Nominal clad temperature	657 °F (347 °C)
Nominal fuel central temperature	4,140 °F (2,282 °C)
Radial peaking factor (variation in power density)	1.5
Axial peaking factor	1.7
Design fuel burnup	32,000 MWd/Te (heavy metal); varies
Fresh fuel assay	3.2% ^{235}U (less in initial load)
Spent fuel assay (design)	0.9% ^{235}U, 0.6% $^{239,241}Pu$
Refueling sequence	One-third of the fuel per year
Refueling time	17 day (minimum)

a. PWRs now being licensed have a 17 × 17 assembly array, with thinner rods totaling 50,952. Other specifications may be slightly changed.

Source: Taken primarily from Westinghouse Electric Corp. specifications.

to the steam generators. From there, the coolant is recirculated to the core by large primary coolant pumps. The main elements of the primary coolant system are shown in Figure 5-5.

The pressure in the primary system is maintained at about 2250 psi (15.5 MPa), preventing the formation of steam. Instead, steam is raised in a secondary system by allowing heat to flow from the high-pressure primary coolant to the lower pressure secondary fluid. This heat transfer occurs through the walls of large numbers of tubes through which the primary coolant circulates in the steam generators. After the steam has passed through separators to remove water droplets,

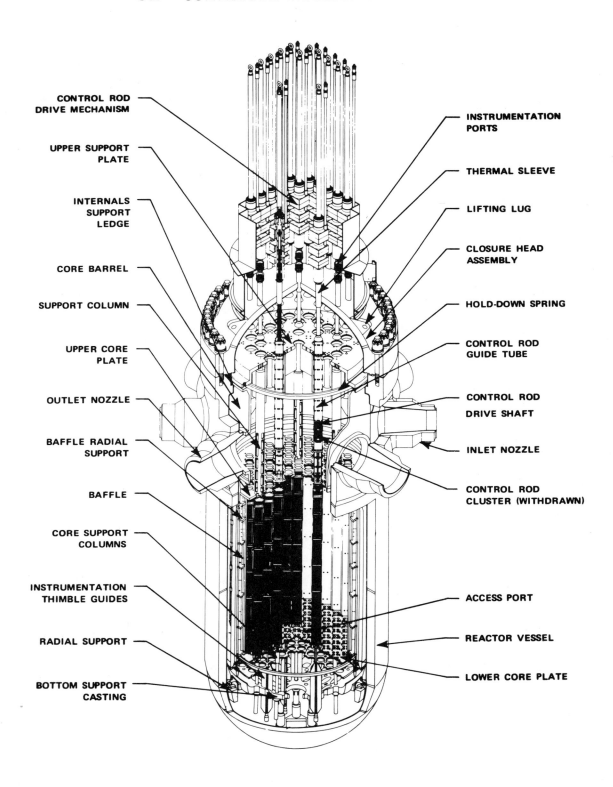

CONTROL ROD
DRIVE MECHANISM

UPPER SUPPORT
PLATE

INTERNALS
SUPPORT
LEDGE

CORE BARREL

SUPPORT COLUMN

UPPER CORE
PLATE

OUTLET NOZZLE

BAFFLE RADIAL
SUPPORT

BAFFLE

CORE SUPPORT
COLUMNS

INSTRUMENTATION
THIMBLE GUIDES

RADIAL SUPPORT

BOTTOM SUPPORT
CASTING

INSTRUMENTATION
PORTS

THERMAL SLEEVE

LIFTING LUG

CLOSURE HEAD
ASSEMBLY

HOLD-DOWN SPRING

CONTROL ROD
GUIDE TUBE

CONTROL ROD
DRIVE SHAFT

INLET NOZZLE

CONTROL ROD
CLUSTER (WITHDRAWN)

ACCESS PORT

REACTOR VESSEL

LOWER CORE PLATE

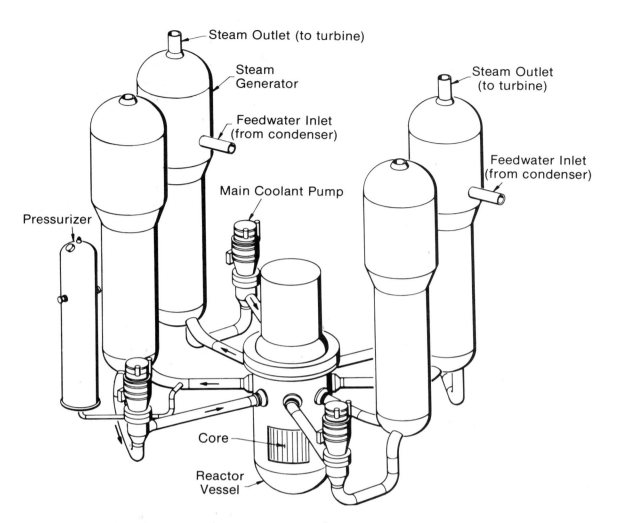

Figure 5-5. ARRANGEMENT OF THE PRIMARY SYSTEM FOR A WESTINGHOUSE PWR.
The primary system constitutes the nuclear steam supply system for a PWR plant. In the four-
loop arrangement shown in the figure, each loop has its own steam generator and coolant pump.
A pressurizer is connected to one of the loops. The primary coolant enters and leaves the
steam generator from the bottom; one of the U-tubes in the generator is shown in Figure 5-8.
(Figure reproduced from WASH-1250.)

Figure 5-4. PRESSURIZED-WATER REACTOR VESSEL AND INTERNALS. (At Left)
The core of a pressurized-water reactor is contained in a large steel vessel through which coolant
flows. After passing into an inlet nozzle, the water flows down between the core barrel and the
vessel wall, until it reaches the plenum beneath the core; there it turns upward to flow through
the core and out one of the outlet nozzles to the steam generators. The top of the reactor
vessel, which is removable for refueling, supports mechanisms for driving control rods. (Figure
courtesy of Westinghouse Electric Corp.)

thereby reducing its moisture content to less than 1%, it proceeds to the turbo-generator for the production of electricity. After condensation, it returns as liquid to the steam generators. The overall thermal efficiency of a PWR is about 32%. In the steam generators, the primary coolant passes only once through a single tube (i.e., the steam generators are "once through"), which is ordinarily either U-shaped or straight. A large PWR may have four external circuits, indicated schematically in Figure 5-5, each with its own steam generator and pump. As seen in Figure 5-6, this arrangement may vary from one manufacturer to another.

Since maintenance of the pressure near the design value is crucial (to avoid the formation of steam in the primary coolant, on the one hand, and rupture of the primary circuit, on the other), a PWR system also includes a "pressurizer," as shown in Figures 5-5 and 5-6, connected to the "hot" leg of one of the steam generator circuits. The pressurizer volume is occupied partly by water and partly by steam; it has heaters for boiling water and sprayers for condensing steam, as needed, to keep the pressure within specified operating limits.

AUXILIARY SYSTEMS

It is useful to mention the systems that support the main reactor systems and that, in addition, are sometimes intimately connected with the safety systems discussed in the next section. These include particularly the systems for controlling the chemistry and volume of the primary reactor coolant and the decay heat removal system.

The chemistry and volume (C & V) control system provides water for the primary coolant system and reduces the concentration of corrosion and fission products in the coolant, as well as adjusting the boric acid concentration. When the reactor is operating, the system functions by continuously bleeding water from the primary coolant system, passing it through demineralizers and into a volume control tank. Liquid supplied to the primary coolant system is some combination of fluids from this tank, from a fresh demineralized water supply, from the boric acid tanks, and from chemicals needed to maintain coolant chemistry within specifications. The C & V system operates in conjunction with the pressurizer to maintain the proper coolant pressure and volume under normal operation. The system may also maintain the proper concentrations of dissolved gases, particularly hydrogen in the coolant. In connection with this function, the C & V system is a source of gas that must be handled by the gaseous waste processing system; the gaseous waste system provides for storage of gas and, in some cases, ultimate return, if necessary, to the reactor system. A liquid waste processing system whose primary purpose is to process liquids from various drain systems may also be connected with the C & V system; when the liquid may contain tritium, such as the primary coolant does, it may be demineralized and returned to the C & V system. The configuration of the C & V system, or its equivalent, and its connection with the primary coolant system and the waste processing systems can vary significantly from one reactor to another. Figures 3-3 and 3-4 provide one example of liquid and gaseous waste control systems.

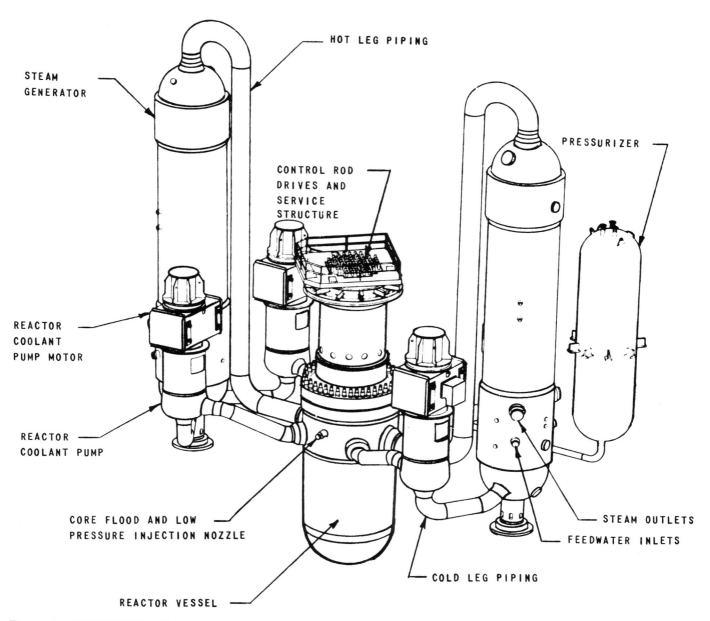

Figure 5-6. ALTERNATIVE ARRANGEMENT FOR A PWR PRIMARY SYSTEM.
This PWR system has two outlet nozzles, each leading to a steam generator. The outlet of each generator is connected with two coolant pumps, each of which is connected with an inlet nozzle at the reactor vessel. These steam generators use vertical tubes, rather than the U-tube design of figure 5-5. (Figure courtesy of Babcock & Wilcox Co.)

The residual heat removal (RHR) system removes decay heat from the primary coolant system during plant shutdown. The system consists primarily of heat exchangers and pumps. At the initial stages of shutdown, heat is still removed by the steam generators, and the resulting steam is discharged directly to the condenser, bypassing the turbine. When the reactor coolant has dropped in temperature and, even more significantly, in pressure, the RHR system is turned on. The cooling function of the steam generators is then removed; one of the reactor coolant pumps continues to operate for a time to ensure uniform residual cooldown. The heat removal system may also be used in conjunction with the emergency injection systems discussed in the next section.

In addition to these specific auxiliary systems, a PWR has numerous other auxiliary systems which provide basic services for the major systems. These include systems for cooling specific components, for providing power (even in emergency situations), and for controlling, via complex electrical networks with either manual or automatic supervision, the functions of the basic systems. Although we do not devote attention here to these numerous systems, they must be adequate to constitute a basis for economic and safe operation of nuclear power plants.

SAFETY SYSTEMS

A number of important safety features are added to the basic reactor system in order to minimize the danger from reactor accidents. The immediate safety function following any abnormality is to shut down rapidly (i.e., to "trip" or "scram") the chain reaction. This is accomplished by the shutdown control rods described earlier. In the event that the abnormality continues to the extent of rupturing the primary system or otherwise reducing coolant inventory, emergency injection systems are available to provide continued cooling of the core. Finally, in the event that fuel melting occurs, the containment building and its subsystems are available to minimize the amount of radioactivity that escapes into the general environment.

Before proceeding to a discussion of the emergency core cooling systems, it is worth noting that the components of both the primary coolant system and the various ECC systems are enclosed by the containment building. Such a building is shown schematically in Figure 5-7. This structure is steel-lined reinforced concrete, designed to withstand the overpressure expected if all the primary coolant were released in an accident. Sprays and cooling systems (such as the relatively new ice condenser system of Figure 5-7) are available for washing released radioactivity out of the containment atmosphere and for cooling the internal atmosphere, thereby keeping the pressure below the containment design pressure. At the initial phases of a severe accident, the containment interior is isolated from the outside world. The basic purpose of the containment system, including its spray and cooling functions, is to minimize the amount of released radioactivity that escapes to the external environment. The basic design criterion is the dose limitation specified by 10 CFR 100 (see Chapter 4).

Meeting these criteria depends, however, on successful operation during emergencies of various systems associated with the reactor. Of primary interest is

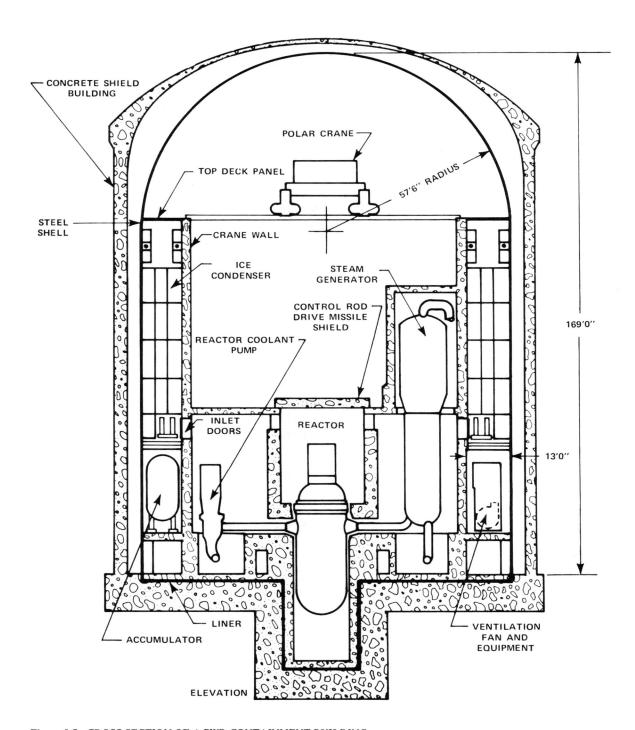

Figure 5-7. CROSS-SECTION OF A PWR CONTAINMENT BUILDING.
The containment building has the entire primary system, as well as various safety systems, in its interior. The building itself is concrete, with a steel shell inside. The safety systems within the building include emergency core cooling systems (note the accumulator), pressure control systems (one form of which may be the ice condenser indicated), and ventilation equipment. (Figure courtesy of Westinghouse Electric Corp.)

Figure 5-8. PWR EMERGENCY CORE
COOLING SYSTEMS.
Several systems are available for supplying
coolant to the core in the event that the
primary system fails. These include a
passive accumulator system, as well as
active injector systems. The effect of a
break in the cold leg of one of the primary
loops is indicated. Note that the core
coolant flow can reverse and that coolant
from other loops can bypass the core.

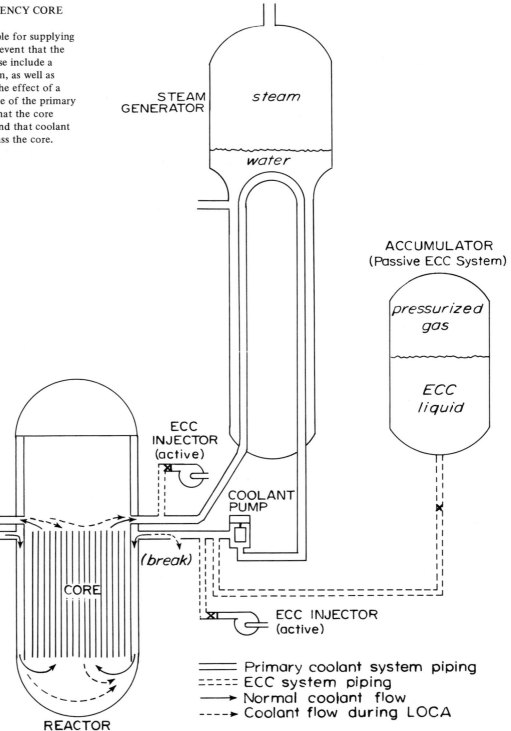

STEAM
GENERATOR   *steam*

water

ACCUMULATOR
(Passive ECC System)

*pressurized
gas*

*ECC
liquid*

ECC
INJECTOR
(active)

COOLANT
PUMP

(break)

CORE

ECC INJECTOR
(active)

REACTOR

——— Primary coolant system piping
----- ECC system piping
——→ Normal coolant flow
----→ Coolant flow during LOCA

the behavior of the systems to be called upon during a loss-of-coolant accident. Such an event can vary greatly in degree, and the several ECC systems are intended to cope with a broad range of accidents, ranging from minor leakage from a small pipebreak to a rapid loss of coolant (blowdown) arising from a complete shear of a main coolant return line ("cold leg") in one of the coolant loops. The only LOCA that these systems are not designed to cope with is a catastrophic rupture of the reactor vessel, in which case there is no system that holds water.

Not surprisingly, there should be little difficulty in dealing with a small to intermediate break. It is, rather, the large breaks whose consequences are most difficult to control. Figure 5-8 indicates schematically the major features of interest in a major cold leg break. Whereas coolant normally flows down the annulus between the core barrel and the reactor vessel, then up through the core, and out to the steam generator, the fluid in the core can reverse direction to flow up the barrel and out the broken leg. Indeed, coolant from other loops can bypass the core to escape out the break. The coolant inventory can be exhausted very rapidly, and any ameliorating action must be massive and rapid. Accordingly, the first system to respond is a passive system, consisting of accumulators which are isolated from the primary system by check-valves that open as soon as the primary system pressure drops much below 1000 psi (7 MPa). Each accumulator has about 1000 cubic feet (28 m^3) of liquid, and each reactor system has two or more units. The accumulators act with no delay, and inject fluid either into the cold legs (as shown in Figure 5-1) or into the reactor vessel. Of course, for the case of a cold leg break with cold leg injection, one of the units would be ineffective. It is also conceivable that other units could be ineffective, as would be the case if fluid injected bypassed the core to escape through the cold leg break.

In any case, this accumulator is rapidly exhausted. Long-term cooling would be provided by two active low-pressure injection systems (LPIS), which pump fluid, each at about 3000 gallons per minute (190 liter/s) into either hot or cold legs, or both. These systems require about 20 seconds to become operative; and it is assumed, in accident analysis, that one of the two systems would be effective.

Finally, for small breaks that do not greatly reduce the pressure, two high-pressure injection systems (HPIS) provide makeup water at relatively low rates (about 400 gallons per minute or 25 liter/s). This water is usually injected into a hot or cold leg. However, the HPIS and LPIS water is injected into the reactor vessel in some designs. The source of water for the active injection systems is typically the volume control tanks and the refueling water storage tanks.

It is the emergency core cooling systems whose operation is uncertain and generates much heated controversy. As noted in Chapter 4, transient conditions during a large LOCA are so difficult to model that, for licensing purposes, a "conservative" model and associated criteria are specified. The differences between such a model and one that is "realistic," but very uncertain, is illustrated in Figure 5-9, showing fuel clad temperatures during a large LOCA. The "conservative" model yields high temperatures, presumed by the Nuclear Regulatory Commission to be an upper limit, for use as a criterion for protecting the public.

The existence alone of emergency systems is not sufficient to limit the course of an accident, even assuming the systems are designed adequately. In addition, the overall reactor system must be arranged to ensure that necessary safety systems operate when required. For this reason, individual systems are duplicated as noted above, and their control and power supplies (for active systems) are independent of each other and of the main reactor systems. Unintended dependencies between systems can reduce the overall dependability of emergency response and can, of course, introduce imponderables into an assessment of the risk from nuclear plant accidents. This question of redundancy and independence also arises in connection with those portions of the main reactor systems that would be used during an accident. For example, of the four primary coolant pumps in a large PWR, each is generally large enough to provide alone for sufficient coolant flow for removal of decay heat after shutdown.

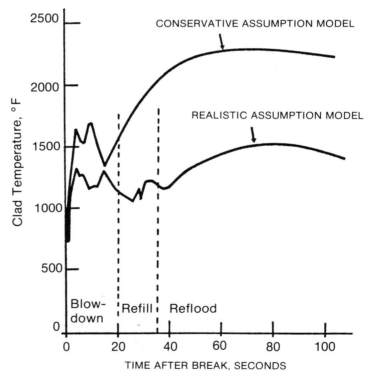

Figure 5-9. SCHEMATIC CALCULATED FUEL CLAD TEMPERATURES FOR A PWR LOCA.
System conditions during an accident may be calculated using both "conservative" and "realistic" models, the first to put an effective limit on the severity of the accident, the second to yield the best available prediction for what will happen. The figure indicates how the models can differ in the results calculated for clad temperatures during the course of a PWR accident. (Figure reproduced from WASH-1250.)

TABLE 5-2.

Approximate Pressurized-Water Reactor Neutronics
(start of life)

Approximately 2.0 fast neutrons are produced following the absorption
of 1 neutron by ^{235}U and have the following fate:

0.6[a]	Captured by ^{238}U (largely in the resonance region, leading to ^{239}Pu production)
1	Absorbed by ^{235}U (of which 0.8 result in fissions)
0.1	Absorbed by water
0.1	Absorbed by structural material and fission product poisons
0.2	Absorbed by control poisons
2.0	

a. The conversion ratio is thus 0.6.

NEUTRONICS, FUEL UTILIZATION, AND REACTOR OPERATION

It is typical of light-water reactors, as they operate at present, that the conversion ratio, the ratio of fissile material produced to that destroyed, is about 0.6. Roughly speaking, for each slow neutron absorbed by ^{235}U, about 2.0 fast neutrons are produced.[1] These are rapidly slowed to thermal energies by the water moderator, but in the process a substantial number are captured by ^{238}U resonances. Of the neutrons that reach thermal energies, some are still captured by ^{238}U, but most are captured by ^{235}U, water, structural materials, fission product poisons, and control poisons. Table 5-2 indicates these results for a PWR just after initial fueling. Note that the ratio of ^{238}U captures (yielding ^{239}Pu) to ^{235}U absorptions (destroying ^{235}U) is about 0.6. As the reactor runs, fission product poisons build up, the amount of fissile material decreases slightly, and the amount of control decreases, so that the tabulated neutron absorptions change slightly. However, the conversion ratio does not change drastically, even though the types of fissile and fertile material will change. (For example, initially the only fissile material is ^{235}U, but reactor operation builds up an inventory of ^{239}Pu and other isotopes.) Note also that the reactor has a large amount of control at startup. Were it possible to reduce this, the conversion ratio would rise. (See, for example, discussion of CANDU, Chapter 7, and of the light-water breeder reactor, Chapter 14.)

In an important sense, the difference between the conversion ratio and 1 is an indicator of resource use. This difference, 1–0.6, is approximately 0.4 for LWRs, indicating a substantial deficit. However, the extent to which fissile resources are used involves other factors, such as whether the fuel reaches its design "burnup" and whether material in the spent fuel is reprocessed. As was noted in Chapter 2, a PWR would require that about 4100 tons of U_3O_8 be supplied to the fuel cycle for its use. Most of this supply is directly associated with the deficit caused by the low

[1] About 0.1 of these result from net neutron production from fission of ^{235}U by fast neutrons.

conversion ratio. Only a small percentage is needed to produce the initial fuel load. If uranium and plutonium are not recycled, this uranium requirement rises by about 50%.

However, all these requirements are based on the assumption that, on the average, the nominal amount of energy is extracted from the fuel rods. The design average burnup[2] of about 32,000 megawatt-days (thermal) per metric ton (MWd/Te) corresponds to a plant capacity factor of about 80%, providing the plant is refueled as scheduled. (Capacity factor is the ratio of actual electrical energy produced to the output if the plant operates continuously at 100% of rated power.) This is substantially higher than the 60% or so that has recently been achieved. If, in spite of relatively low average output for a plant, refueling proceeds on schedule, not as much energy will have been extracted from the fuel. This can represent a net loss of resources if the fuel is not reprocessed and fissile material recycled. Most PWRs have been constructed on the presumption that refueling would occur once yearly, in a low demand period, but possible losses of energy value may cause reexamination of such strict scheduling. The initial design has typically required a burnup of about 10,000 MWd/Te between refuelings, but lower burnup, for whatever reason, may warrant postponement of refueling.

Various factors may cause such low burnup. Most notable from the public health point of view are shutdowns due to difficulties with safety related equipment. Often, though, shutdowns occur because of other maintenance needs. The refueling shutdown (see below) takes a substantial amount of time. On the other hand, low capacity factor (and burnup) may arise from operating the plant at lower than nominal output. Reduced output may occur as a result of safety-related deratings, or as a result of reduced electrical demand. Normally, a nuclear power plant is designed as a base-load[3] unit, so that it ordinarily runs at full output, but as the portion of nuclear units in a utility grid grows, these units may more often be required to follow demand. PWRs can alter load easily enough, using control rods, to accommodate themselves to such needs. However, use in such a mode will reduce the capacity factor.

When refueling occurs, the reactor is unavailable for a substantial period, a minimum of two weeks. During this period, plant workers commonly receive a substantial portion of their annual radiation dose. Standard practice in controlling this dose is to flood the region around the reactor vessel in water, so that fuel is handled underwater. Fuel is moved by a conveyer between an opening in the side of the containment and the point where it is lifted over the edge of the open pressure vessel. In a PWR, the entire head (see Figure 5-4) is removed, along with the control rod drives. A portion of the inner core is removed, assemblies from the periphery are moved into this region, and fresh fuel is added at the periphery. A subsequent period for reconnecting and testing contributes substantially to the shutdown time of about two weeks.

---

[2] Because burnup will vary from one fuel rod to another, the fuel rods are designed to withstand higher burnup to make the average figure of 32,000 MWd/Te possible.

[3] Base-load plants are operated continuously in order to supply the minimum demand on a utility's grid. However, the utility will experience both daily and seasonal increases above this demand; these increases are met by peak and intermediate load units, usually fossil-fuel fired plants.

Bibliography — Chapter Five

"NTIS" indicates report is available from: National Technical Information Service, U.S. Department of Commerce, 5285 Port Royal Road, Springfield, VA 22161.

APS-1975. H. W. Lewis et al., "Report to the American Physical Society by the Study Group on Light-Water Reactor Safety," *Reviews of Modern Physics*, vol. 47, supplement no. 1, p. S1 (1975).
> Treats basic PWR systems, emphasizing safety systems and the U.S. reactor safety research program.

"Coastal Effects of Offshore Energy Systems: An Assessment of Oil and Gas Systems, Deepwater Ports, and Nuclear Power Plants off the Coast of New Jersey and Delaware," U.S. Congress Office of Technology Assessment, U.S. Government Printing Office (November 1976).
> Includes a brief description of proposed floating PWR power plants.

"Environmental Statement Related to Construction of Koshkonong Nuclear Plant," (draft), U.S. NRC report NUREG-0079 (August 1976).
> Summary of environmental aspects of a proposed PWR nuclear power plant.

ERDA-1541. "Final Environmental Statement, Light Water Breeder Reactor Program, Commercial Application of LWBR Technology," 5 vols., U.S. ERDA report ERDA-1541 (June 1976) (NTIS).
> Environmental statement for the light-water breeder, essentially a PWR.

"Final Environmental Statement Related to the Operation of Rancho Seco Nuclear Generating Station," U.S. AEC report (March 1973).
> Summary of environmental aspects of a PWR power plant.

GESMO. "Final Generic Environmental Statement on the Use of Recycle Plutonium in Mixed Oxide Fuel in Light Water Cooled Reactors: Health, Safety, and Environment," 5 vols., U.S. NRC report NUREG-0002 (August 1976) (NTIS).
> Describes environmental implications of recycling plutonium in LWRs.

McPherson, G. D., "Results of the First Three Nonnuclear Tests in the LOFT Facility," *Nuclear Safety*, vol. 18, p. 306 (1977).
> Reviews recently completed tests at the LOFT facility, a small-scale PWR, instrumented for LOCA measurements.

Steam, Its Generation and Use, 28th ed. (Babcock & Wilcox, 1972).
> Describes use of steam for power generation, including electrical generation via PWR power plants.

"Systems Summary of a Westinghouse Pressurized Water Reactor Power Plant," Westinghouse Electric Corporation report (1971).
> Describes systems and operation of a PWR power plant.

WASH-1250. "The Safety of Nuclear Power Plants (Light-Water Cooled) and Related Facilities," U.S. AEC report WASH-1250 (July 1973) (NTIS).
> Includes a summary of PWR systems.

WASH-1400. "Reactor Safety Study: An Assessment of Accident Risks in U.S. Commercial Nuclear Power Plants," 9 vols., U.S. NRC report WASH-1400, NUREG-75/014 (October 1975) (NTIS).
> Study of light-water reactor safety, actually calculating the risk from reactor accidents.

Ybarrondo, L. J., Solbrig, C. W., and Isbin, H. S., "The 'Calculated' Loss-of-Coolant Accident: A Review," American Institute of Chemical Engineers Monograph Series, no. 7 (1972).
> Includes a brief description of PWR systems, particularly those related to LOCAs.

CHAPTER SIX

Boiling-Water Reactors

ABOUT A THIRD of light-water reactors operating or under construction in the United States are boiling-water reactors. The distinguishing characteristic of a BWR is that the reactor vessel itself serves as the boiler of the nuclear steam supply system. In fact, the reactor vessel and associated equipment is the NSSS, as suggested in Figure 6-1. This vessel is by far the major component in the reactor building, and the steam it produces passes directly to the turbogenerator. The reactor building also contains emergency core cooling equipment, a major part of which is the pressure suppression pool which is — as suggested in Figure 6-2 — an integral part of the containment structure. As noted later in the chapter, earlier BWRs utilized a somewhat different containment and pressure suppression system. All the commercial BWRs sold in the United States have been designed and built by General Electric.

Several types of reactors that use boiling water in pressure tubes have been considered, designed, or built. In a sense, they are similar to the CANDU, described in Chapter 7, which uses pressure tubes and separates the coolant and moderator. The CANDU itself can be designed to use boiling light water as its coolant. The British steam-generating heavy-water reactor (Chapter 7) has such a system. Finally, the principal reactor type now being constructed in the Soviet Union uses a boiling-water pressure tube design, but with carbon moderator.

BASIC BWR SYSTEM

A boiling-water reactor core consists of a large number of fuel assemblies, each a square array as indicated in Figure 6-3. Although many BWRs use a 7 × 7 array, the most recent model (BWR/6) uses an 8 × 8 array, with thinner fuel rods; the cross-sectional size of the newer fuel bundle is therefore similar to the 7 × 7 array. The fuel pin is very similar to that discussed in Chapter 5 (Figure 5-2), with an active length of at least 12 feet (3.6 m). Unlike the typical PWR fuel bundle,

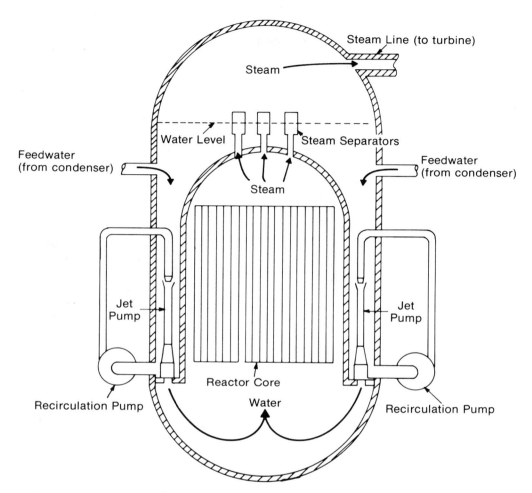

Figure 6-1. SCHEMATIC ARRANGEMENT OF A BOILING-WATER REACTOR.
In a boiling-water reactor, the steam for driving the turbogenerator is formed in the reactor
vessel itself. Water passes through the core, forming steam which proceeds to the turbine.
Water that is still liquid is recirculated in the vessel through the action of "jet pumps" which
surround the core (see text). (Figure reproduced from WASH-1250.)

that of the BWR has an outer sheath (fuel channel) which constrains the flow of
water in the assembly. An orifice at the bottom of the bundle then strongly determines the flow rate for a given assembly. The structural stability of the assembly is
supplied by upper and lower tie plates, together with tie rods which take eight of the
64 array positions in an 8 × 8 assembly. (See the 4-assembly cross-section of Figure
6-3.) In addition, the assembly has several fuel rod spacers. Assemblies may also
contain water rods (rods with water rather than UO_2), providing moderator within
the bundle. A large BWR contains 764 assemblies, with 40 or 50 thousand fuel
rods, and about 180 tons (160 metric tons) of UO_2.

The cross-shaped object around which the four bundles are arranged in Figure

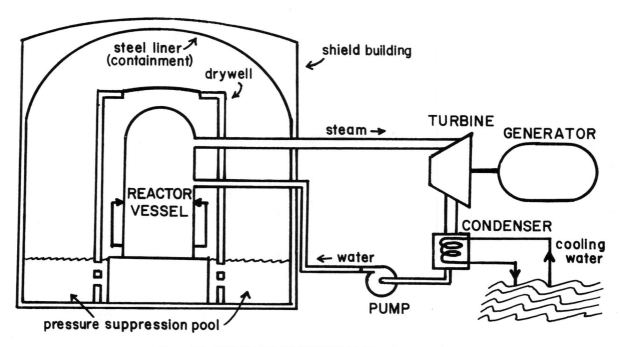

Figure 6-2. SCHEMATIC OF BOILING–WATER REACTOR POWER PLANT.
Steam from a BWR reactor vessel flows to the turbogenerator, after which it is condensed and
returned as feedwater to the reactor vessel. The reactor vessel is contained in a dry well which,
in turn, is within a reactor building.

6-3 is the cruciform control element used in BWRs. This element actually contains numerous boron-carbide-filled rods, one quarter in each of the blades shown. The cruciform rods are driven from the bottom of the reactor. These rods serve for both reactivity control and power flattening. The reactivity control includes long-term regulation and prompt shutdown ("scram"). Power flattening is needed in particular because, as the coolant rises through the core, it boils, resulting in lower coolant densities, and therefore poorer neutron moderation and lower power densities in the upper portion of the core. Burnable poisons are present as an oxide of gadolinium ("gadolinia") mixed into several of the fuel rods per bundle; this poison is present in all fresh fuel and is completely depleted during one year of operation. The reactor is also controlled by the recirculation rate (see below).

At refueling, assemblies are removed from the central core region and replaced by assemblies from the periphery. Fresh fuel is then added to the periphery of the core. Fresh fuel has an average enrichment of 2.4 to 3.0%. Within an assembly, the enrichment will vary, with lower enrichment fuel in the corners and near the water gaps; it is in these regions that neutrons are more effective because they are better thermalized. A major goal, as usual, is to achieve a relatively flat power distribution. The average power generation density in the core is about 51 kW/liter.

The coolant flow rate is about 105 million pounds per hour (13 Mg/s); the feed-water temperature is about 376 °F (191 °C), and water exiting the core is about 550 °F (288 °C), maintaining the clad temperature below 600 °F (316 °C). (See Table 6-1 for reactor parameters.)

The core and associated equipment are contained in a large, steel reactor

TABLE 6-1.
Representative Characteristics of Boiling-Water Reactors

Core thermal power	3,579 MWth
Plant efficiency	34%
Plant electrical output (nominal)	1,220 MWe
Core diameter	193 in (4.9 m)
Core (or fuel rod) active length	150 in (3.8 m)
Core weight (fuel assemblies)	524,000 lb (238 Mg)
Core power density	54 kW/liter
Cladding material	Zircaloy-2
Cladding diameter (OD)	0.483 in (1.23 cm)
Cladding thickness	0.032 in (0.81 mm)
Fuel material	UO_2
Pellet diameter	0.410 in (1.04 cm)
Pellet height	0.41 in (1.04 cm)
Assembly array	8 × 8, with fuel channel enclosing array
Number of assemblies	748
Total number of fuel rods	46,376
Control rod type	"Cruciform" control rods inserted from the bottom between sets of four assemblies.
Number of control rods	177
Total amount of fuel (UO_2)	342,000 lb (155 Mg)
Fuel/coolant ratio	1/2.7, blades out; 1/2.5, blades in (cold)
Coolant	Water (two phase)
Total coolant flow rate	104×10^6 lb/hr (13 Mg/sec)
Coolant pressure	1,040 psia (7.0 MPa)
Coolant temperature (steam system design)	551 °F (288 °C)
Feed water temperature	420 °F (216 °C)
Average coolant exit quality (percent steam weight)	14.7%
Average clad temperature	579 °F (304 °C)
Maximum fuel central temperature	3,330 °F (1,832 °C)
Average volumetric fuel temperature	1,130 °F (610 °C)
Axial peaking factor	1.4 approx.
Design fuel burnup	28,400 MWd/Te
Fresh fuel assay	Average 2.8% ^{235}U (initial core: 1.7-2.1% ave.)
Spent fuel assay	0.8% ^{235}U, 0.6% $^{239,241}Pu$
Refueling sequence	Approximately one-fourth of the fuel per year to one-third per 18 months
Refueling time	188 hrs @ 100% efficiency
Vessel wall thickness min/max	5.7 in/6.46 in (14.5 cm/16.4 cm)
Vessel material	Manganese-molybdenum-nickel steel internally clad with 1/8 in austenitic stainless steel
Vessel diameter (ID)	19 ft 10 in (6.0 m)
Vessel height	71 ft (22 m)
Vessel weight (including head)	1,950,000 lb (884,500 Kg)

Source: General Electric Co. specifications.

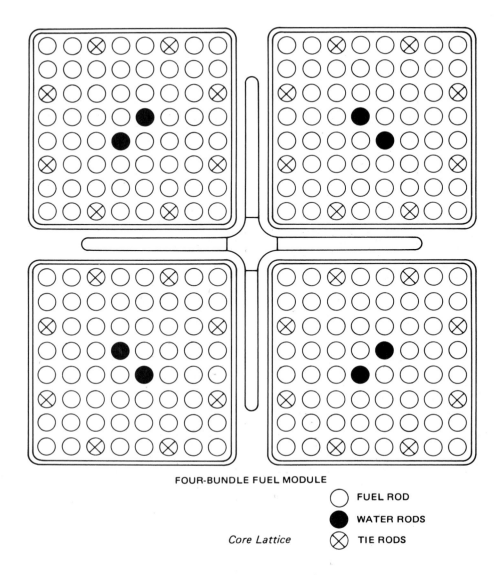

FOUR-BUNDLE FUEL MODULE

○ FUEL ROD

● WATER RODS

Core Lattice   ⊗ TIE RODS

Figure 6-3. BOILING–WATER REACTOR CORE COMPONENTS.

CORE LATTICE

The basic module of a BWR core is a set of four fuel bundles, with a control assembly at the point where they meet. Note that some of the positions in the fuel assemblies are taken by tie rods and others are occupied by water rods, which serve to flatten the power distribution in the assembly.

FUEL ASSEMBLY

A BWR fuel assembly consists of a square array of fuel rods, held together by upper and lower tie plates and interim spacers, and surrounded by a fuel channel. The bottom of the assembly serves to regulate the flow through the assembly.

CONTROL ROD

The BWR control rod is a four-bladed assembly containing neutron absorber rods. This assembly is driven from the bottom of the reactor vessel. (Figure courtesy of General Electric Co.)

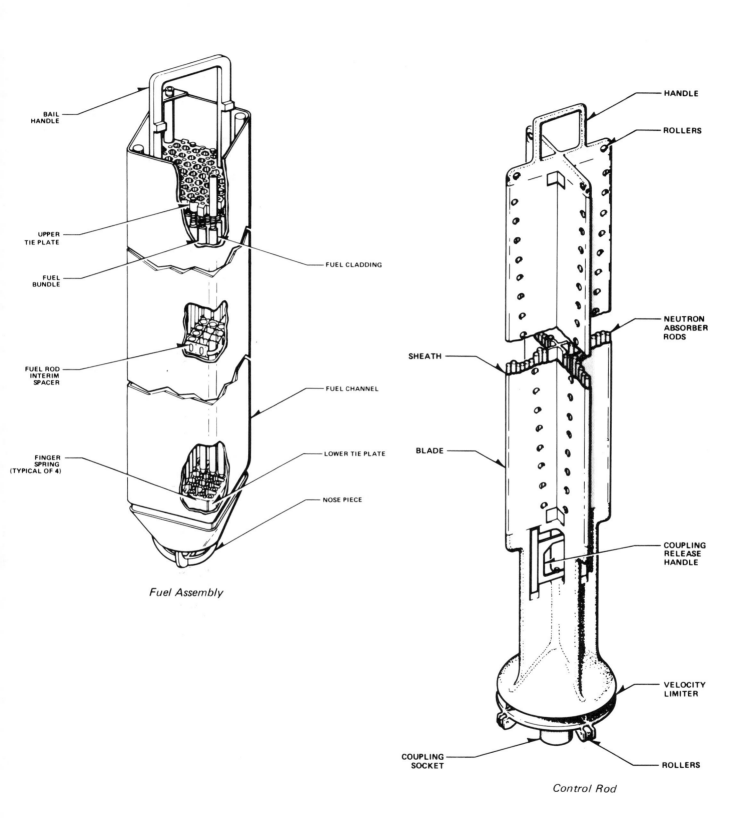

BAIL
HANDLE

UPPER
TIE PLATE

FUEL
BUNDLE

FUEL ROD
INTERIM
SPACER

FINGER
SPRING
(TYPICAL OF 4)

FUEL CLADDING

FUEL CHANNEL

LOWER TIE PLATE

NOSE PIECE

Fuel Assembly

HANDLE

ROLLERS

NEUTRON
ABSORBER
RODS

SHEATH

BLADE

COUPLING
RELEASE
HANDLE

VELOCITY
LIMITER

COUPLING
SOCKET

ROLLERS

Control Rod

vessel (see Figure 6-4). In addition to the fuel assemblies, the other nuclear components of major interest are the control rods, which are mounted on the bottom of the reactor vessel, with drives below. The top head of the vessel is removable for refueling and contains no large equipment. Above the core are steam separators and dryers, comparable to devices in a PWR steam generator. The vessel containing all this equipment is very large, about 72 feet (22 m) in height and 21 feet (6 m) in diameter for a large BWR. It is made of carbon steel, 6 to 7 inches (16 cm) thick and all but the top, which comes into contact only with high-quality steam, is clad with 1/8 in. (0.3 cm) stainless steel. The vessel can withstand pressures greater than 1000 psi (7 MPa) at operating temperatures.

As suggested by Figure 6-1, the water in the vessel boils as it rises through the core. The BWR system is maintained at a pressure of about 1000 psi (7 MPa), at which pressure water boils at a temperature of 545 °F (285 °C). Of course, not all the water passing through the core is vaporized. About 13% (by weight) of the fluid leaving the core is steam. The remainder is recirculated down an annulus formed between the core "shroud" and the reactor vessel, to the plenum beneath the core. The fluid then passes up again through the core.

The steam generated is separated from the remaining liquid by a structure of steam separators which are positioned above the core, at the interface between the predominately liquid and gaseous phases. Steam from the separators then passes through a dryer assembly which removes moisture. The dried steam proceeds out of the vessel, through the drywell wall and reactor building (see below), to the turbogenerator. (Unlike the PWR system, the steam from a BWR — coming as it does directly from the core — is radioactive, primarily because of the presence of nitrogen 16, an isotope with a 7-second half-life.) Steam from the turbines is condensed and returned as feedwater to the reactor vessel, where it joins the flow recirculating to the bottom of the vessel. The thermal efficiency of a BWR is about 33%.

As we have indicated, most of the coolant recirculates within the reactor vessel, rather than in an external loop. This flow is pumped by a series of jet pumps in the annulus outside the core shroud. The jet pumps are basically reactor inlet nozzles for two external recirculation systems, each with a recirculation pump and associated valves and piping (see Figure 6-5). About one third of the core flow is taken from the reactor vessel and pumped through the manifold and jet pumps, thereby driving the annular flow as a whole. The water then turns upwards into the individually orificed fuel assemblies, as discussed above. The recirculation rate serves as one of the control systems. If the flow rate is decreased, a greater percentage of the water rising through the core is changed to steam, so that neutrons are less effectively moderated. The reaction rate and core power therefore tend to drop.

Figure 6-4. REACTOR ASSEMBLY OF A BWR POWER PLANT.
The reactor vessel of a BWR contains not only the core assembly but also devices for separating and drying steam. This steam is generated as the coolant flows up through the core. As the remaining liquid returns along the outside of the core, a portion of it is drawn off to the recirculation system and returned through the jet pumps, which thereby cause the bulk recirculation within the reactor vessel. (Figure courtesy of General Electric Co.)

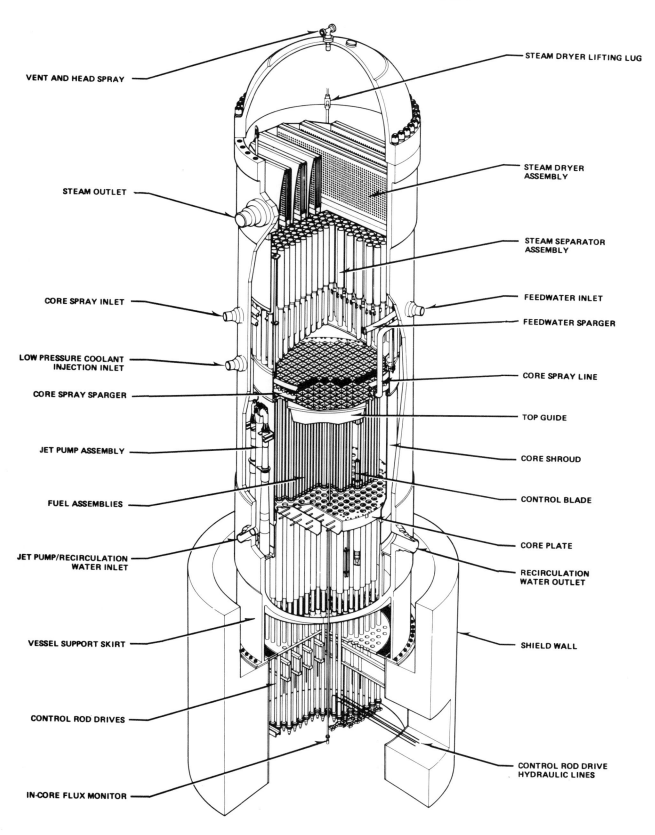

VENT AND HEAD SPRAY

STEAM DRYER LIFTING LUG

STEAM OUTLET

STEAM DRYER ASSEMBLY

STEAM SEPARATOR ASSEMBLY

CORE SPRAY INLET

FEEDWATER INLET

FEEDWATER SPARGER

LOW PRESSURE COOLANT INJECTION INLET

CORE SPRAY SPARGER

CORE SPRAY LINE

TOP GUIDE

JET PUMP ASSEMBLY

CORE SHROUD

FUEL ASSEMBLIES

CONTROL BLADE

CORE PLATE

JET PUMP/RECIRCULATION WATER INLET

RECIRCULATION WATER OUTLET

VESSEL SUPPORT SKIRT

SHIELD WALL

CONTROL ROD DRIVES

CONTROL ROD DRIVE HYDRAULIC LINES

IN-CORE FLUX MONITOR

Figure 6-5. JET PUMP RECIRCU-
LATION SYSTEM.
Water from a recirculation outlet is
pumped back into the reactor vessel
via several jet pumps, thereby driving
the coolant within the reactor vessel.
(Figure courtesy of General Electric
Co.)

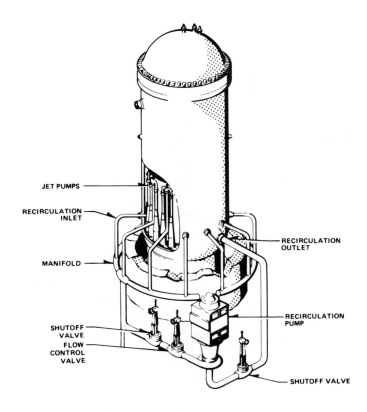

AUXILIARY SYSTEMS

Like the other water-cooled reactors, the BWR has systems for controlling water chemistry (and volume) and for removing decay heat. In the following brief discussion, aspects of these systems that are peculiar to the BWR are emphasized.

The coolant cleanup system removes fission products, corrosion products, and other impurities from a stream of water that is drawn off via the recirculation pump line and returns via the feedwater line. Cleaning is accomplished by filter-demineralizer units. In addition to performing a cleaning function, this system is also used to remove the excess water volume caused by lowering of the coolant density (due to boiling) as the reactor is brought up to power.

Decay heat removal after reactor shutdown is accomplished by a residual heat removal (RHR) system that is largely a part of the emergency core cooling system discussed below. Decay heat removal is ordinarily accomplished by drawing water from the recirculation line, cooling it in a heat exchanger, and returning it to the feedwater line. Emergency functions of the RHR system are discussed below.

Of the various other BWR systems, most provide basic services such as power, component cooling, and system control. The only system that is peculiar to the BWR is the system for cleaning and cooling the fuel storage and containment pools. The containment pools are a distinctive aspect of BWRs and are discussed in the context of safety design.

SAFETY SYSTEMS

The basic containment configuration for BWRs is shown in Figure 6-6, a schematic drawing of the Mark III containment and shield building. The reactor vessel and immediately associated equipment, such as the recirculation system and the pressure relief valves on the main steam lines, are enclosed in a drywell, which seals the reactor from the rest of the reactor building. The atmosphere in the drywell is in contact with a pressure suppression pool which forms an annulus around the drywell. In recent designs (Figure 6-6), the drywell is a concrete structure, and the suppression pool is on the floor of the reactor building between the containment liner and the drywell wall. The pool connects to the interior of the drywell through horizontal vents, but is prevented from covering the drywell floor by a "weir" wall; an upper containment pool sits atop the drywell. In earlier designs (Figure 6-7), the drywell consists of a steel primary containment, and the pressure suppression pool (with large numbers of downcomer tubes) is contained in a large torus connected to the drywell by several large vent pipes. In either case, blowdown of the reactor coolant inventory into the drywell tends to raise the pressure, thus forcing fluid into the pressure suppression pool. There steam is condensed, thus controlling the pressure increase.

In current designs, a steel containment shell surrounds all the equipment of the reactor building. This containment provides a sealed barrier against radioactive releases and is designed to withstand temperatures and pressures that could be caused by a loss-of-coolant accident. Surrounding the containment is the reactor building itself, a reinforced concrete structure which further limits radioactive releases and also protects the containment from external agents (weather, missiles).

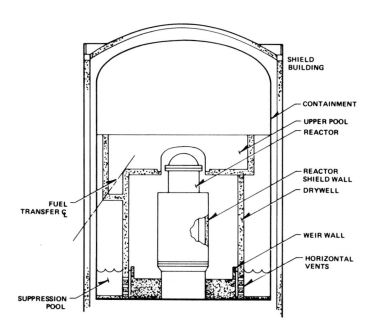

Figure 6-6. BWR MARK III CONTAINMENT AND SHIELD BUILDING.
The BWR reactor vessel is contained within a concrete drywell, which in turn is contained within a reactor building with a steel containment. The drywell is surrounded by a pressure suppression pool, which communicates with the drywell interior through horizontal vents. There is, in addition, a pool above the reactor. (Figure courtesy of General Electric Co.)

Figure 6-7. BWR MARK I PRIMARY CONTAINMENT.
In older versions of the BWR, the reactor vessel is enclosed in a dry well which communicates, via vent pipes and a downcomer system, with a pressure suppression pool contained in a large torus. This entire structure is contained in a reactor building. (Figure reproduced from WASH-1250.)

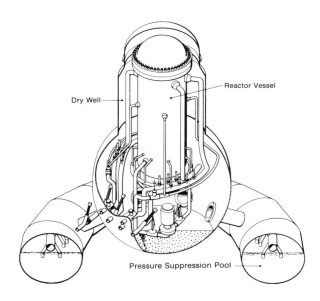

The annulus between the building and containment is maintained at negative pressure to serve as a collector of radioactivity during accident conditions. The atmosphere of this annulus is filtered to collect suspended radioactive materials.

Numerous systems are available for controlling abnormalities. In the event that control rods cannot be inserted, liquid neutron absorber (containing a boron compound) may be injected into the reactor to shut down the chain reaction. Heat removal systems are available for cooling the core in the event the drywell is isolated from the main cooling systems. Closely related to the heat removal systems are injection systems for coping with decreases in coolant inventory.

Both abnormalities associated with the turbine system and actual loss of coolant accidents can lead to closing of the steam lines and feedwater line, effectively isolating the reactor vessel within the drywell. Whenever the vessel is isolated, and indeed whenever feedwater is lost, a reactor core isolation cooling system is available to maintain coolant inventory by pumping water into the reactor via connections in the pressure vessel head. This system operates at normal pressures and initially draws water from tanks that store condensate from the turbine, from condensate from the residual heat removal system, or, if necessary, from the suppression pool.

A network of systems performs specific ECC functions to cope with LOCAs. (See Figure 6-9.) These all depend on signals indicating low water level in the pressure vessel or high pressure in the drywell, or both. The systems include low-pressure injection, utilization of the RHR system, and high- and low-pressure core spray systems. The high-pressure core spray is intended to lower the pressure within the pressure vessel and provide makeup water in the event of a LOCA. In the event the core is uncovered, the spray can directly cool the fuel assemblies. Water is taken from the condensate tanks and from the suppression pool. On the other hand, should it become necessary to use the low-pressure systems, the vessel must be depressurized. This can be accomplished by opening relief valves to blow down the vessel contents into the drywell (and hence the suppression pool). Once this is done, the low-pressure core spray may be used to cool the fuel assemblies (drawing water from the suppression pool) or RHR low-pressure injection (again from the suppression pool) may be initiated, or both. The RHR system may also be used

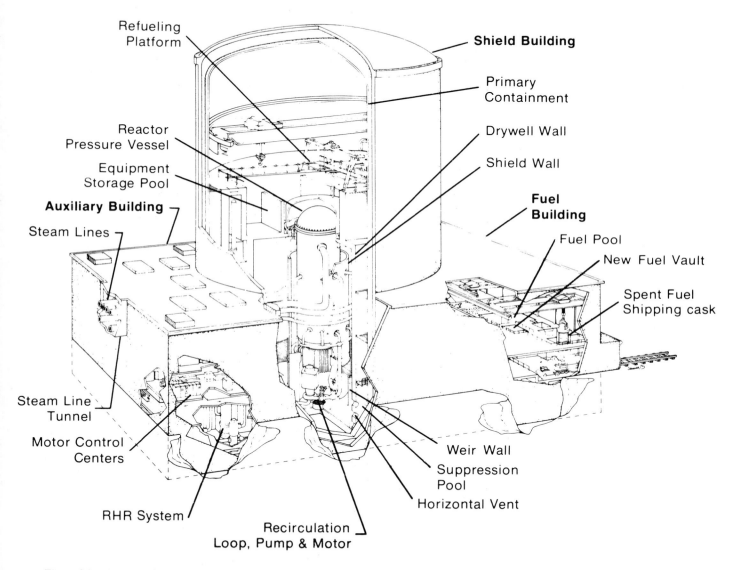

Figure 6-8. BWR REACTOR BUILDINGS.
Directly connected with the containment and shield building of a BWR are a fuel building and an auxiliary building. The turbine building is not shown. (Figure courtesy of General Electric Co.)

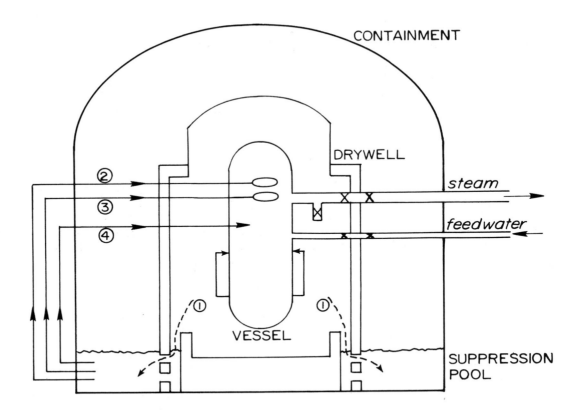

Emergency core cooling functions:

① overpressure injection into pressure suppression pool
② high pressure core spray
③ low pressure core spray
④ low pressure coolant injection

(✗ valves)

Figure 6-9. BWR EMERGENCY CORE COOLING SYSTEMS.
Several systems are available for supplying coolant to the core in the event that the basic BWR systems fail. The basic system for condensing and collecting coolant, thereby limiting drywell pressure, is the passive pressure suppression pool. In addition, active systems provide for high- and low-pressure core spray and for low-pressure coolant injection.

simply to cool the suppression pool. (Two other functions of the RHR are to provide decay heat removal under ordinary shutdown conditions and, when necessary, to supplement the cooling system for the spent fuel pool and the upper containment pool.)

The various ECC systems are thus designed to cool the core adequately under any conditions that are apt to occur. Ultimately, the water supply for any of the injection or spray systems is the suppression pool. This is also where reactor coolant losses should flow, so that a closed loop should exist. Thus the pressure suppression pool acts, not only to condense steam, thereby controlling containment pressure, but also to provide an emergency coolant reservoir.

BWR ECC systems have not been as controversial as those of the PWR, partly because the performance of the BWR spray systems, located above the core, is easier to analyze than the PWR ECC systems. However, it is interesting to note that the Reactor Safety Study (see Chapter 4) concluded that, within the uncertainties of their results, the risk from BWRs and PWRs were not markedly dissimilar. Moreover, the 1975 experience at Brown's Ferry, where burned cables led to a situation whereby coolant inventory was slowly being lost, showed that unexpected circumstances can circumvent multiple safety systems.

In many ways, the remarks at the end of the discussion of PWR safety systems (Chapter 5) apply equally well to BWRs. Both conservative and realistic models of emergency core cooling function exist, and as in any reactor system, great attention is given to assuring redundancy and independence of safety systems.

NEUTRONICS, FUEL UTILIZATION, AND REACTOR OPERATION

The neutronics and fuel utilization of a BWR are grossly similar to those of a PWR, for which the reader is referred to Chapter 5. As in a PWR, the actual burnup achieved by a BWR depends on how the reactor is operated. A BWR has a somewhat unusual capability for varying output to meet demand in that alteration of the coolant flow rate changes the reaction rate. This method of load following is not available to other types of reactors. A BWR also differs from a PWR in that it has a larger volume of fuel available for a given rated output. As a result, not only is the BWR power density lower, but the residence time of the fuel may be longer, particularly if comparable burnups are achieved. Since this appears to be the case (the BWR is designed for 27,500 MWd/Te versus the PWR's 32,000), refueling may not have to occur as often. General Electric does, in fact, cite one possible refueling sequence as replacing about one third of the core every 18 months. Moreover, on the newer systems (BWR/6 in Mark III containment) a refueling time of one week is specified. Older BWRs take a longer time.

In general, refueling entails opening the top of the drywell and removing the vessel head, steam dryers, and steam separators. The reactor well area is filled with water, and fresh and spent fuel bundles are exchanged in the upper containment pool area. A refueling tube connects this area with the fuel storage areas in the fuel building attached to the shield building (see Figure 6-8).

Bibliography — Chapter Six

"NTIS" indicates report is available from: National Technical Information Service, U.S. Department of Commerce, 5285 Port Royal Road, Springfield, VA 22161.

APS-1975. H. W. Lewis et al., "Report to the American Physical Society by the Study Group on Light-Water Reactor Safety," *Reviews of Modern Physics*, vol. 47, supplement no. 1, p. S1 (1975).

Treats basic BWR systems, particularly their safety systems and related research.

"Final Environmental Statement, Perry Nuclear Power Plant," U.S. AEC report (April 1974).

Summary of environmental aspects of a BWR power plant.

"General Description of a Boiling Water Reactor," General Electric Company report (March 1976).

Summary of BWR systems and their operation.

GESMO. "Final Generic Environmental Statement on the Use of Recycle Plutonium in Mixed Oxide Fuel in Light-Water Cooled Reactors: Health, Safety, and Environment," 5 vols., U.S. NRC report NUREG-0002 (August 1976) (NTIS).

Describes environmental implications of recycling plutonium in LWRS.

WASH-1250. "The Safety of Nuclear Power Plants (Light-Water Cooled) and Related Facilities," U.S. AEC report WASH-1250 (July 1973) (NTIS).

Includes a summary of BWR systems.

WASH-1400. "Reactor Safety Study: An Assessment of Accident Risks in U.S. Commercial Nuclear Power Plants," 9 vols., U.S. NRC report WASH-1400, NUREG-75/014 (October 1975) (NTIS).

Study of LWR safety, actually calculating the risk from reactor accidents.

Ybarrondo, L. J., Solbrig, C. W., and Isbin, H. S., "The 'Calculated' Loss-of-Coolant Accident: A Review,' American Institute of Chemical Engineers Monograph Series, no. 7 (1972).

Includes a brief description of BWR systems, including those related to LOCAs.

CHAPTER SEVEN

Heavy-Water Reactors

AN ALTERNATIVE to using ordinary water as the moderator and coolant of a thermal reactor is to choose "heavy" water for one or both of these purposes. Because heavy water absorbs fewer neutrons than ordinary water, heavy water moderated reactors (HWR) can be designed with natural uranium (0.7% ^{235}U) as the fuel. Moreover, because of the lower absorption and because the heavy water is a somewhat less effective moderator, it is feasible and advantageous to have larger separations between fuel bundles than in an LWR. This leads to the possibility of having individually cooled fuel channels, one bundle thick, with heavy-water moderator surrounding the channels. This is the basic configuration of commerical HWRs. These HWRs typically utilize a pressurized (as opposed to boiling) primary coolant system, so that a schematic of the reactor coolant and generating system is identical to that of a pressurized-water reactor (see Figures 1-4 and 5-1) except that the primary coolant may be heavy water. This is the case in the reactor now being marketed by Atomic Energy of Canada, Limited (AECL), the CANDU, for "Canadian deuterium-uranium" reactor. Most of the discussion in this chapter focuses on the CANDU, particularly its newer versions.

Although current CANDUs use heavy water, not only as the moderator, but also as the coolant, other cooling fluids are possible. Two that have been seriously considered, both in Canada and elsewhere, are light water and organic coolant. Light water is much less expensive than heavy water. Organic materials can operate at higher temperatures, thereby improving the thermal efficiency of the power plant.

In recent years, a significant portion of the British nuclear program has been directed to development of a "steam generating heavy-water reactor" (SGHWR). The SGHWR uses light-water coolant in vertical pressure tubes, which are immersed in heavy water moderator. The coolant is permitted to boil, and steam is separated in a steam drum, from which it goes to the turbine, as in a boiling-water reactor.

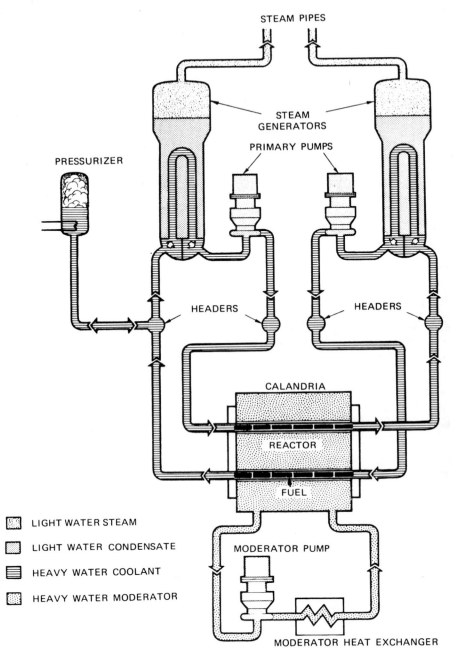

Figure 7-1. SCHEMATIC FLOW DIAGRAM FOR A CANDU POWER PLANT.
Present CANDU systems are essentially pressurized-water reactors. Individual fuel channels pass through a calandria, which contains heavy water moderator with its own circulation system. Heavy water coolant, on the other hand, flows through the fuel channels and raises steam (from ordinary water) in the steam generators. (Figure courtesy of Atomic Energy of Canada Ltd.)

The system uses slightly enriched uranium as its fuel. Britain has been developing the SGHWR as the basis of its nuclear power system. However, this choice is being reconsidered.

BASIC HWR SYSTEM

The important distinction, of course, between an LWR and an HWR is that the moderator of the latter is heavy water. In both of the reactor types cited, CANDU and SGHWR, a lattice of fuel channels is immersed in a pool of heavy-water moderator. The coolant passes through the channels and may be heavy water, light water, or some other fluid. In the case of the current CANDU, it is heavy water. A schematic diagram of the CANDU reactor and coolant system is given in Figure 7-1. Note that the fluid in the secondary loops, which drive the turbogenerators, is light water.

The fuel of a CANDU is similar to that of an LWR in that fuel pellets of uranium dioxide are sealed into Zircaloy-clad fuel pins, which are bound into bundles. A 600-MWe CANDU would have about 4500 bundles, containing about 100 tons (90 Mg) of uranium dioxide. However, in the case of the current CANDU, the uranium has only the natural concentration of ^{235}U, 0.7%. Moreover, the pins are arranged into bundles, shown in Figure 7-2, that are somewhat smaller and

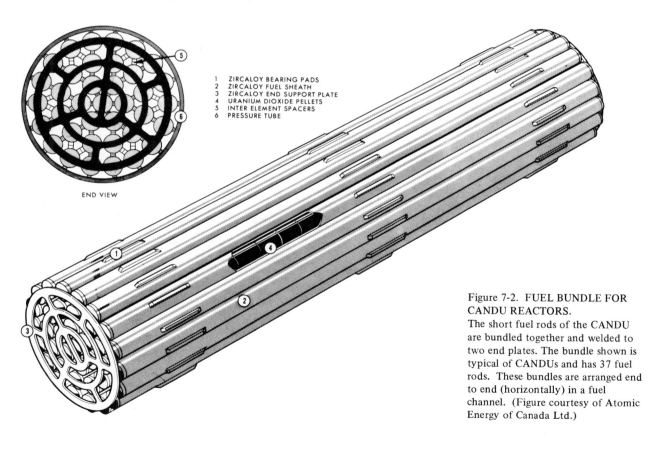

1 ZIRCALOY BEARING PADS
2 ZIRCALOY FUEL SHEATH
3 ZIRCALOY END SUPPORT PLATE
4 URANIUM DIOXIDE PELLETS
5 INTER ELEMENT SPACERS
6 PRESSURE TUBE

END VIEW

Figure 7-2. FUEL BUNDLE FOR CANDU REACTORS.
The short fuel rods of the CANDU are bundled together and welded to two end plates. The bundle shown is typical of CANDUs and has 37 fuel rods. These bundles are arranged end to end (horizontally) in a fuel channel. (Figure courtesy of Atomic Energy of Canada Ltd.)

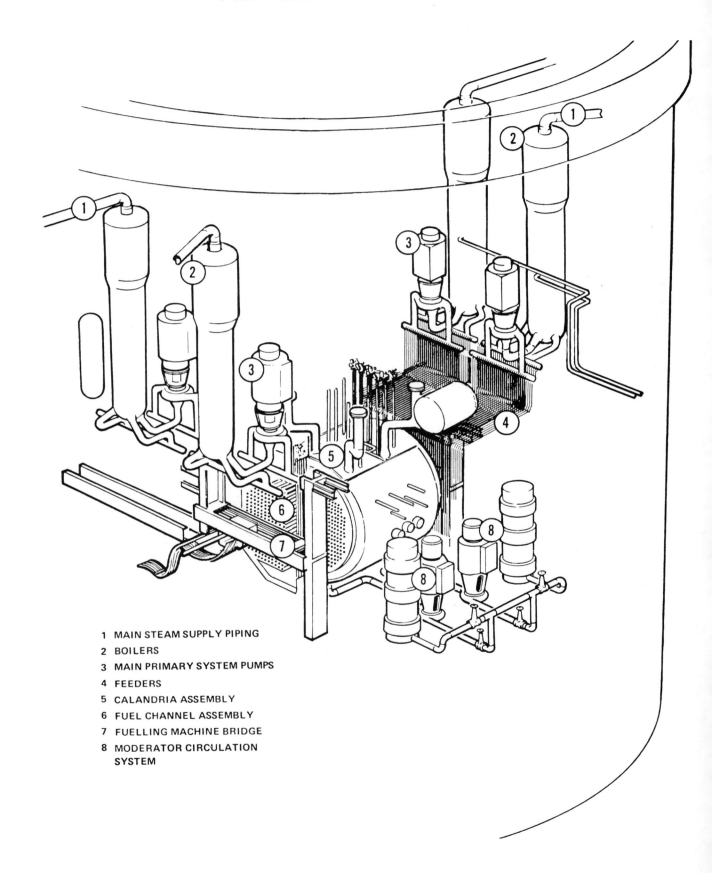

1 MAIN STEAM SUPPLY PIPING
2 BOILERS
3 MAIN PRIMARY SYSTEM PUMPS
4 FEEDERS
5 CALANDRIA ASSEMBLY
6 FUEL CHANNEL ASSEMBLY
7 FUELLING MACHINE BRIDGE
8 MODERATOR CIRCULATION
   SYSTEM

simpler than those of LWRs. These bundles do not have hardware for maintaining the core configuration, a function that is performed by the fuel channels. Instead, the bundles and channels are designed for on-line refueling. On the average, about 15 bundles are replaced per day of operation, without shutting down the reactor. This has some advantage, perhaps, in that no refueling shutdown is necessary. However, its most important consequence from the point of view of reactor design is that relatively little neutron absorber is necessary during reactor operation, because there are no large swings in fissile content and fission product poisons during the fuel cycle. This leads to a higher conversion ratio and, under some conditions, to significantly improved resource utilization (see end of this chapter).

Figure 7-1 shows only two of the fuel channels. In an actual reactor, there are hundreds of channels, each with a row of fuel bundles arranged end to end. These fuel channels pass horizontally through a lattice of tubes which is part of a "calandria" which contains the moderator (see Figure 7-3). This moderator, heavy water, is maintained at near atmospheric pressure, so that this reactor system does not require fabrication of a large pressure vessel. The calandria is moderate in size, a cylinder about 25 feet (7.6 m) in diameter and 25 feet (7.6 m) long, made with stainless steel walls about 1 in. (2.5 cm) thick, and ends about 2 in. (5 cm) thick. The calandria tubes are made of Zircaloy. The moderator in the calandria has its own cooling system (including two pumps and two heat exchangers) which maintains moderator temperature at about 160 °F (70 °C). (See Table 7-1 for representative parameters.) During operation, the vault containing the calandria is filled with water.

The primary coolant system is similar to that of a PWR except that the pressure vessel is replaced by a lattice of hundreds of individual pressure tubes, each with a feeder at either end leading to headers at the pumps and steam generators. Individual pressure tubes may be opened during reactor operation for refueling. The tubes are fabricated from an alloy of zirconium and there is a gas space between the pressure tube and the surrounding calandria tube. The heavy-water coolant is maintained at a pressure of about 1500 psi (10 MPa) and, in passing through the pressure tubes, reaches a temperature of 590 °F (310 °C), below the boiling point at that pressure. The primary coolant flow pattern is relatively simple: coolant from a primary pump passes through a distribution header to the individual tubes, goes once through the reactor, through the header at the steam generator, and through the U-tube steam generator to the primary pump. The flow rate (600 MWe CANDU) is about 60 million pounds per hour (7.6 Mg/s). In the present CANDU (called a "pressurized heavy water reactor" for obvious reasons), there are four steam generators and pumps, paired to achieve the flow patterns shown in Figure

---

Figure 7-3. PRIMARY SYSTEM FOR A CANDU REACTOR.
Numerous fuel channels pass through the CANDU calandria. Each is connected via its own pipes to the headers at a primary coolant pump and at a steam generator. There is, in addition, a circulation and cooling system for the moderator contained in the calandria. (Figure courtesy of Atomic Energy of Canada Ltd.)

TABLE 7-1.
Representative Characteristics of a CANDU Reactor[a]

Core thermal power	2,140 MWth
Plant efficiency	28%[a]
Plant electrical output	600 MWe
Core diameter	248 in (6.3 m)
Core length	234 in (5.9 m)
Core weight (fuel bundles)	240,000 lb (109 Mg)
Core power density	12 kW/liter (core average within calandria)
Cladding material	Zircaloy
Cladding diameter (OD)	0.515 in (1.31 cm)
Cladding thickness	0.016 in (0.04 cm)
Fuel material	UO_2
Pellet diameter	0.478 in (1.21 cm)
Fuel bundle array	37 rods, arranged in concentric circles
Array diameter (OD)	4 in (10 cm)
Total number of bundles	4,560
Total number of fuel rods	168,720
Total amount of fuel (UO_2)	210,000 lb (95 Mg)
Control rod types	Variable neutron absorbers (light-water compartments), adjustable absorbers (such as stainless steel); shutdown by absorbing rods or poison injection
Number of control rods or compartments	From 4 to 21 of each type of absorber
Coolant	Heavy water (liquid, plus some gas phase), >95% D_2O
Total coolant flow rate	60×10^6 lb/hr (7.6 Mg/s)
Coolant pressure (entrance to channel)	1,602 psi (11.1 MPa)
Coolant pressure (exit of channel)	1,493 psi (10.3 MPa)
Coolant temperature (entrance)	512 °F (267 °C)
Coolant temperature (exit)	594 °F (312 °C)
Average coolant exit quality	3%
Moderator	Heavy water, 99.75% D_2O (molecular ratio)
Moderator pressure	Approximately atmospheric
Moderator temperature (entrance)	110 °F (43 °C)
Moderator temperature (exit)	160 °F (71 °C)
Total heavy water inventory	1.02×10^6 lb (463 Mg)
Maximum fuel temperature	3,832 °F (2,110 °C)
Maximum clad temperature	684 °F (362 °C)
Axial peaking factor	1.5
Radial peaking factor	1.2
Fuel residence time	470 full-power days
Design fuel burnup	7,000 MWd/Te[a]
Fresh fuel assay	0.71% ^{235}U
Spent fuel assay	0.2% ^{235}U, 0.3% $^{239,241}Pu$
Refueling sequence	On-line, essentially continuous, refueling
Calandria outer diameter	25 ft (7.6 m)
Calandria length	25 ft (7.6 m)
Calandria wall thickness (stainless steel)	1-1/8 in (3 cm) thick walls, 2 in (5 cm) ends
Number of calandria tubes (Zircaloy)	380
Lattice array	Square with 11 in (28 cm) pitch

a. The detailed design varies from one reactor to another. In particular, newer models have slightly different dimensions, somewhat higher fuel burnup and efficiency.

SOURCE: Atomic Energy of Canada Ltd. specifications.

7-1. The system pressure is maintained by a single pressurizer, connected to the headers at two of the steam generators.

The secondary coolant fluid in a CANDU is light water. As in any steam power plant, this steam drives a turbine, is condensed, then returned to the boilers (steam generators) as feedwater. The overall thermal efficiency of a CANDU system is about 29%, significantly lower than that of most commercial nuclear power plants.

Reactivity control is achieved by several systems, including (light) water zone control absorbers, solid absorber rods, and poisons for addition to the moderator. (In some older models, control has been via highly enriched fuel rods, whose withdrawal reduces the reactivity.) In current CANDUs, routine on-line control is accomplished by the zone absorbers, which consist of compartments in the core into which light water, a neutron absorber, can be introduced. In addition, several mechanical control rods (containing cadmium) supplement this control and can be dropped under gravity for quick power reduction. Two banks of about 14 cadmium control rods are available specifically for reactor shutdown. Long-term reactivity control and startup reactivity control, respectively, are provided by neutron absorbing compounds of boron and gadolinium in the moderator. Finally, core power shaping is achieved by stainless steel adjuster rods. In addition, the power distribution can be effectively controlled by the refueling sequence, since only one pressure tube is serviced at a time.

AUXILIARY SYSTEMS

Systems are available for performing important service functions for the main system, including chemistry and volume control and shutdown cooling. These are similar to those for a PWR except for the differences required by the separate moderator and coolant systems.

The moderator cleanup system controls impurities and includes the capability for removing boron and gadolinium neutron poisons. The coolant purification system takes flow from a primary pump outlet and returns it to the pump inlet; the system uses filtering and ion exchange for removing impurities. The coolant volume control system is closely linked with the pressurizer and has enough capacity to handle all changes in coolant volume associated with alterations in power level. Because of the expense of heavy water (about $100/kg), the reactor building contains systems for the collection, purification, and upgrading of heavy water, in order to minimize inventory losses.

Two shutdown cooling systems connect to the reactor inlet and outlet headers, essentially in parallel with the primary pumps and steam generators. As the reactor cools down, these systems, each with a pump and heat exchanger, gradually take over decay cooling. Initially, pumping force through the heat exchangers is provided by the primary pumps, but, as the coolant temperature decreases, shutdown pumps assume this function and the primary pumps and steam generators are isolated.

SAFETY SYSTEMS

Under abnormal conditions, the first action is to shut down the reactor. This is accomplished by gravity drop of the shutdown control rods. For cases where these rods could not be inserted, earlier CANDUs had provision for dumping the moderator out of the calandria into a large tank. In current versions this capability is replaced by a fast-acting system for injecting gadolinium into the moderator.

The CANDU has an emergency core cooling system for controlling loss-of-coolant accidents. Should a reactor coolant system rupture, valves close to isolate the intact system, and light water from a storage tank (dousing tank) built into the roof of the containment system is injected into the ruptured system. Heat is initially rejected through the steam generators. As the dousing tank is emptied, water is recovered from the bottom of the reactor building, passed through a heat exchanger, and reinjected into the ruptured system. The moderator in the calandria provides some independent heat capacity, with heat removal provided by the heat exchangers in the moderator circulation system.

A design with many pressure tubes has an advantage in that gross failure of the pressure vessel is not possible. On the other hand, a large LOCA can still occur; for example, one of the headers could be ruptured. However, the other independent coolant loop would presumably still be intact. Furthermore, in the extreme case where all the coolant was lost and the ECC system failed, although the fuel and pressure tubes would be severely damaged, the moderator could carry off enough heat to prevent gross melting.

The containment structure (Figure 7-4) is a prestressed concrete building with a plastic liner. Its subsystems include a spray system and air coolers for reducing the building pressure. In some designs, the containment atmosphere is ordinarily at negative pressure with respect to the external environment.

NEUTRONICS, FUEL UTILIZATION, AND
REACTOR OPERATION

Heavy-water reactors have an advantage over LWRs in that relatively few neutrons are lost to absorption by the moderator. CANDUs in particular have the advantage of on-line refueling. These two effects are the most significant factors in permitting design of a reactor with a conversion ratio (CR) that approaches 0.75 to 0.80. The fact that (1 – CR) is only 0.20 to 0.25 means that operation of a CANDU requires significantly less resource depletion than an LWR, for which (1 – CR) is approximately 0.4. However, this advantage is fully realized only if fissile material in the spent fuel is recovered. If not, the resource utilization of a CANDU is comparable to, or somewhat poorer than, that of an LWR with fissile recycle. The

Figure 7-4. CANDU REACTOR BUILDING.
A reactor building contains the entire primary system of a CANDU, as well as various safety-related systems. The building itself is concrete with a plastic liner. (Figure courtesy of Atomic Energy of Canada Ltd.)

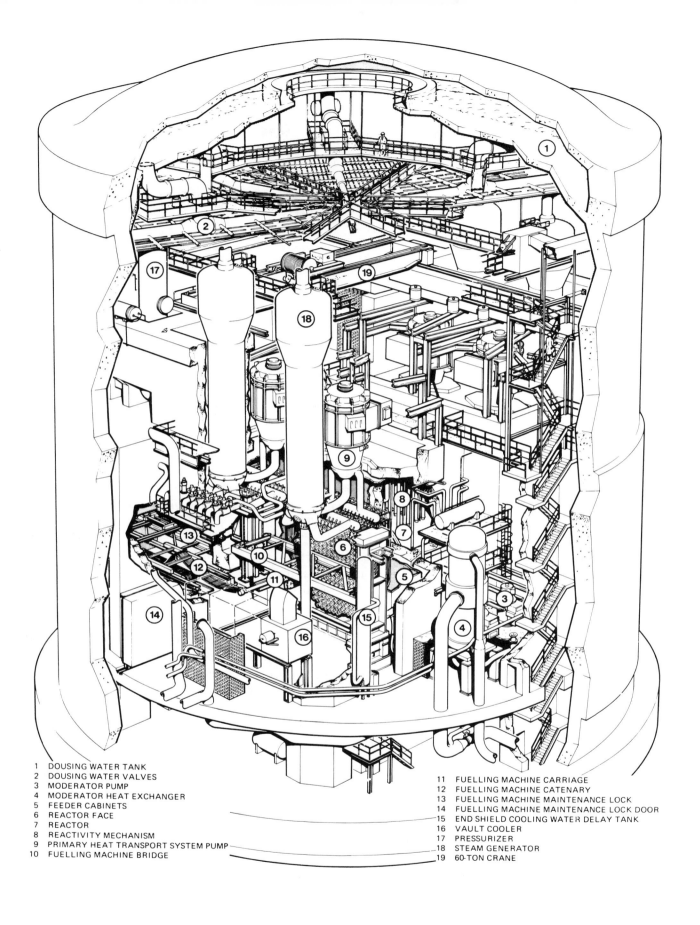

1 DOUSING WATER TANK
2 DOUSING WATER VALVES
3 MODERATOR PUMP
4 MODERATOR HEAT EXCHANGER
5 FEEDER CABINETS
6 REACTOR FACE
7 REACTOR
8 REACTIVITY MECHANISM
9 PRIMARY HEAT TRANSPORT SYSTEM PUMP
10 FUELLING MACHINE BRIDGE

11 FUELLING MACHINE CARRIAGE
12 FUELLING MACHINE CATENARY
13 FUELLING MACHINE MAINTENANCE LOCK
14 FUELLING MACHINE MAINTENANCE LOCK DOOR
15 END SHIELD COOLING WATER DELAY TANK
16 VAULT COOLER
17 PRESSURIZER
18 STEAM GENERATOR
19 60-TON CRANE

TABLE 7-2.
Approximate CANDU Neutronics (equilibrium cycle)

Approximately 2.1 fast neutrons are produced following the absorption
of 1 neutron by fissile material and have the following fate:

0.79[a]	Captured by fertile material, leading to fissile production
1	Absorbed by fissile material (of which 0.8 result in fission)
0.02	Absorbed by heavy water
0.22	Absorbed by structural and fission products
0.06	Absorbed by other materials, including control poisons
0.04	Lost by leakage
2.13	

a. The conversion ratio is therefore 0.79 for this system. However, this high a ratio has not yet been achieved for the CANDU; 0.70 to 0.75 is typical.

full potential of a CANDU is realized only if it is operated near break-even on a thorium cycle (see Chapter 14).

To indicate the manner in which neutrons are used in a CANDU, Table 7-2 summarizes the neutrons produced as the result of one thermal neutron absorption in fissile material. As in an LWR, about two fast neutrons ultimately result, and their final disposition differs from that in an LWR (Table 5-2) in subtle, but important, ways. Note that the conversion ratio, the ratio of fissile material produced to fissile material destroyed, is 0.79. This is possible largely because, of the 2.1 neutrons resulting from absorption by fissile, less than 0.1 are lost to absorption by moderator and control. (This contrasts with 0.3 for LWRs, as noted in Chapter 5.)

The fissile content of fresh fuel in a CANDU is only 0.7%. Not surprisingly, the design burnup is much less than in LWRs — about 8000 MWd/Te. It is interesting to note, too, that the fissile content of the discharged fuel is about 0.5%, slightly more than half of which is fissile plutonium. Whereas the lifetime uranium commitment to a CANDU (1000 MWe) would be about 4200 tons of U_3O_8[1] on a throwaway fuel cycle (see Table 10-1), this would be reduced by about half were the plutonium to be recycled. However, so much more material must be reprocessed and fabricated that, from an economic point of view, there is much less incentive to recycle plutonium in a CANDU than in an LWR.

The fact that CANDUs are continuously refueled offers a clear advantage in fuel management. The utility is never faced with the decision whether to refuel on schedule even when the fuel has not reached design burnup. Fueling can take place as needed, so that the maximum energy may be extracted from the fuel. In a way, the refueling machine acts as a reactivity control, increasing the fissile content precisely when it is required. The on-line refueling may also reduce outage time, but the extent of such reduction is not clear since, during refueling shutdowns, other types of power plants are also serviced in other ways. A disadvantage of on-line

[1] This assumes a burnup of 9600 MWd/Te, a goal that has not yet been achieved.

refueling is that inspection to monitor diversion of nuclear materials (see Chapter 12) becomes more difficult.

Having mentioned economics above, we might go on to note two other such factors. The fact that CANDUs do not require enriched uranium significantly reduces CANDU fuel cycle costs relative to those of LWRs. However, the need for a million pound heavy water (actually 0.4 Mg/MWe) inventory, mostly at the start of the operation, substantially raises the initial cost of the power plant, so that these two characteristics of the CANDU tend to balance one another.

Bibliography — Chapter Seven

"NTIS" indicates report is available from: National Technical Information Service, U.S. Department of Commerce, 5285 Port Royal Road, Springfield, VA 22161.

AECL-1973. "CANDU Nuclear Power Station," Atomic Energy of Canada Limited publication PP-15 (October 1973).

A general description of the CANDU nuclear power system (superseded by AECL-1976).

AECL-1976. "CANDU 600 Station Design," Atomic Energy of Canada Limited publication PP-28 (May 1976).

A general description of AECL programs, of the development of the CANDU; includes parameters for the current 600 MWe version.

EPRI NP-365. "Study of the Developmental Status and Operational Features of Heavy Water Reactors," Electric Power Research Institute report EPRI NP-365 (February 1977) (NTIS).

Study of the status and features of HWRs, particularly as they may pertain to the U.S. nuclear power program.

Foster, J. S., and Critoph, E., "The Status of the Canadian Nuclear Program and Possible Future Strategies," *Annals of Nuclear Energy,* vol. 2, p. 689 (1975).

Describes Canada's current and future use of CANDU reactors.

McIntyre, H. C., "Natural-Uranium Heavy-Water Reactors," *Scientific American,* vol. 233, p. 17 (October 1975).

An elementary summary of the features of CANDU reactors.

Till, C. E., et al., "A Survey of Considerations Involved in Introducing CANDU Reactors into the United States," Argonne National Laboratory report ANL-76-132 (January 1977) (NTIS).

Considers the issues involved in a decision whether to utilize CANDU-type reactors in the United States, including licensing and economic questions.

CHAPTER EIGHT

Gas-Cooled Thermal Reactors

A N IMPORTANT ALTERNATIVE to hydrogen as the moderator in a thermal reactor is carbon. As noted in Appendix B, carbon, with atomic mass 12, requires more collisions to slow down neutrons than does water (see "slowing down power" on Table B-1), but it also absorbs a smaller proportion of neutrons. Because the "moderating ratio," a measure of the slowing down power relative to the absorption, is even better for carbon than light water, designing a reactor with a relatively large mass of carbon can be very effective neutronically. This approach has been taken in numerous reactor systems, including the earliest reactors, which utilized natural uranium as the fuel. In most cases, the coolant in a carbon-moderated reactor is a gas, such as helium or carbon dioxide, but this is by no means necessary: in many Russian carbon-moderated reactors, the coolant is water confined to pressure tubes; the molten salt breeder reactor (Chapter 14) immerses carbon in a liquid fuel salt.

Several gas-cooled carbon-moderated commercial nuclear power plants have been designed. In Great Britain, a number of carbon dioxide cooled reactors have actually been built; this "advanced gas reactor" (AGR) is sometimes considered as an alternative to the SGHWR (Chapter 7). In the United States, the General Atomic Company has built one 330-MWe gas-cooled reactor, but the larger commercial versions were withdrawn from the market in 1976. A similar reactor, but with a "pebble bed," is being developed in Germany (Chapter 14). Interest in these reactors survives, largely because a high-temperature gas coolant offers the potential for high thermal efficiency, particularly in a direct cycle with a gas turbine, and for industrial process heat production. Moreover, the level of interest has risen in connection with the search for more proliferation resistant nuclear systems (see Chapters 12 and 14).

The reactor offered by General Atomic affords a good opportunity to examine the features of gas-cooled reactors. This reactor, called a "high-temperature gas-cooled reactor" (HTGR), uses helium coolant and a core consisting of stacked

120

carbon blocks with small uranium-thorium fuel regions. The basic heat transfer diagram of this reactor (see Figure 8-1) is similar to that of a PWR, except that the primary system of an HTGR contains helium, not water, and the core consists of stacked carbon blocks, not metal fuel rods. Details are given in the rest of this chapter.

BASIC HTGR SYSTEM

The HTGR differs in two major respects from the reactors described in previous chapters. The fuel/moderator system is radically different, since the fuel consists of uranium and thorium pellets contained in fuel regions of carbon moderator blocks. The primary coolant system is distinctive, both because the coolant is a gas, helium, and because the entire primary coolant system is contained in a large prestressed concrete reactor vessel (PCRV), as indicated by the dashed line on Figure 8-1. The general appearance of the core and the physical layout of the primary system are shown in Figure 8-2.

The HTGR core consists of a massive pile of hexagonal graphite blocks, each containing fueled regions, as well as holes for passage of the pressurized helium gas. The fuel itself consists of highly enriched uranium as the fissile material and thorium as the fertile. These fuels, in the form of the dioxide or carbide, both ceramics, are present as small fuel kernels with ceramic coatings. The two types of pellet, shown in Figure 8-3, have different coatings in order to facilitate separation at reprocessing: the fissile pellets, with uranium enriched to 93% ^{235}U, or with recycled ^{233}U, are coated with pyrolitic carbons and silicon carbide; the fertile

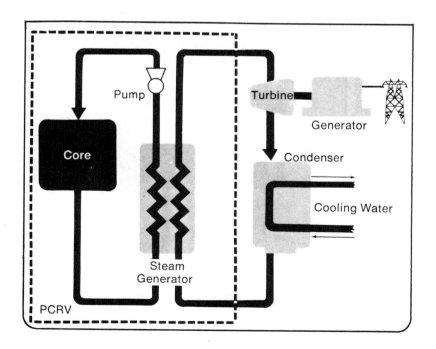

Figure 8-1. SCHEMATIC OF HIGH-TEMPERATURE GAS-COOLED REACTOR POWER PLANT.
The core of an HTGR is mostly carbon, with uranium and thorium fueled regions. Heat from the core is carried off by helium coolant to steam generators. The core, steam generators, helium circulators, and other equipment are contained in a prestressed concrete reactor vessel (PCRV). (Figure reproduced from ERDA-76-107.)

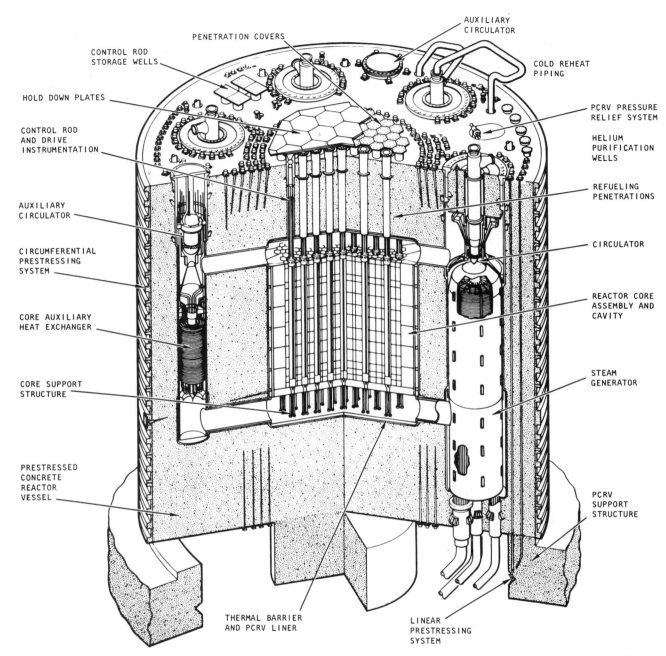

Figure 8-2. HTGR PRESTRESSED CONCRETE REACTOR VESSEL ARRANGEMENT.
The primary system components are contained in a large cylinder of prestressed concrete.
Penetrations exist for refueling, as well as for servicing (and even replacing) various pieces of
equipment. Several primary coolant loops, as well as secondary cooling loops, are contained in
the vessel. (Figure courtesy of General Atomic Co.)

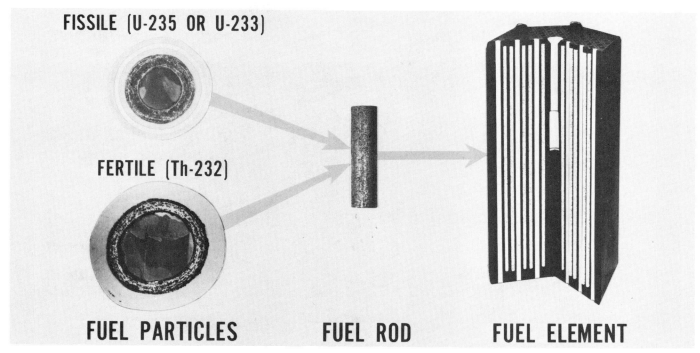

FISSILE (U-235 OR U-233)

FERTILE (Th-232)

FUEL PARTICLES   **FUEL ROD**   **FUEL ELEMENT**

Figure 8-3. HTGR COATED FUEL PARTICLES, ROD, AND ELEMENT.
The HTGR uses two particle types: fissile material is coated with layers of carbon and silicon carbide, fertile material only with carbon. The particles are incorporated into a carbon binder to form a fuel rod, and these are put into fuel elements. (Figure courtesy of General Atomic Co.)

pellets (^{232}Th) are coated with only the carbon. As the reactor runs, fissile ^{233}U builds up in the latter particles. The silicon carbide, because it does not burn, aids in separating the two particle types at reprocessing, where the carbon is burned away.

The fuel particles are incorporated into fuel rods, with graphite as the binder, which are incorporated into the basic block or element (Figure 8-3). These elements are stacked as indicated in Figure 8-4. A basic refueling region consists of a central stack, which has two vertical control rod penetrations, and the adjacent six stacks, without such channels. The PCRV penetration above the central stacks (see Figure 8-2) serves both for refueling and, during operation, for the control drive mechanism. The central stacks also have an additional channel into which boron carbide balls can be poured as a reserve shutdown system. All the fuel elements have holes through which the coolant flows.

The core and other components of the nuclear steam supply system are contained in various cavities of the PCRV (Figure 8-2). Each of the cavities is steel lined to provide a seal and protect the concrete vessel. For detail of the core cavity

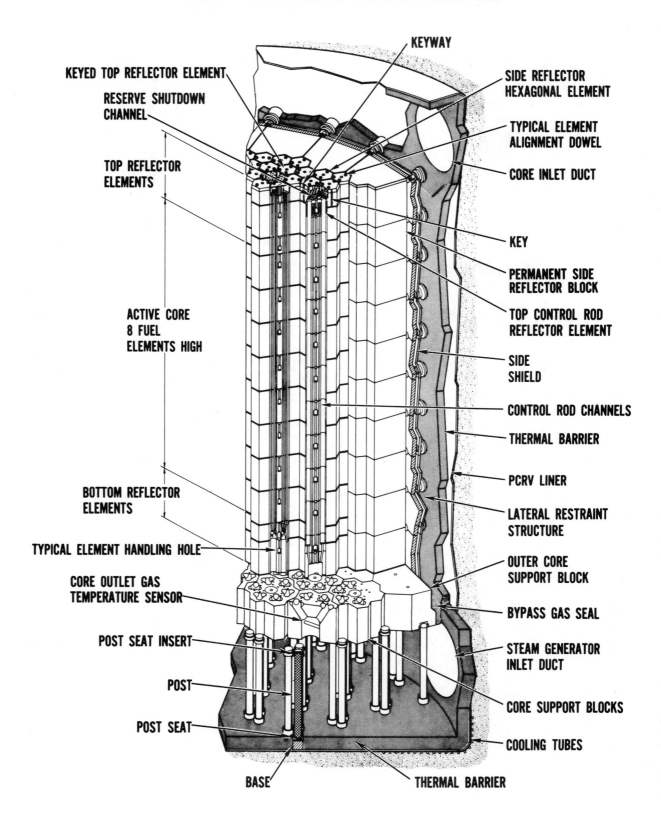

KEYWAY

KEYED TOP REFLECTOR ELEMENT

SIDE REFLECTOR HEXAGONAL ELEMENT

RESERVE SHUTDOWN CHANNEL

TYPICAL ELEMENT ALIGNMENT DOWEL

CORE INLET DUCT

TOP REFLECTOR ELEMENTS

KEY

PERMANENT SIDE REFLECTOR BLOCK

ACTIVE CORE 8 FUEL ELEMENTS HIGH

TOP CONTROL ROD REFLECTOR ELEMENT

SIDE SHIELD

CONTROL ROD CHANNELS

THERMAL BARRIER

PCRV LINER

BOTTOM REFLECTOR ELEMENTS

LATERAL RESTRAINT STRUCTURE

TYPICAL ELEMENT HANDLING HOLE

OUTER CORE SUPPORT BLOCK

CORE OUTLET GAS TEMPERATURE SENSOR

BYPASS GAS SEAL

POST SEAT INSERT

STEAM GENERATOR INLET DUCT

POST

CORE SUPPORT BLOCKS

POST SEAT

COOLING TUBES

BASE

THERMAL BARRIER

lining, see Figure 8-4. The PCRV has penetrations for refueling and control, as noted above, and for piping. In addition, there are removal plugs for servicing, and even replacement, of steam generators, helium circulators, etc. (It has been found, however, at the 330 MWe Fort St. Vrain (Colorado) HTGR, that imperfections in the core lining are difficult to repair.) The vessel is prestressed with vertical steel tendons and wrapped with circumferential cables. The PCRV and its contents are extremely massive, about 100 million pounds (45,000 Mg); and indeed the core itself is more massive, by about an order of magnitude, than the core of an LWR. See Table 8-1 for HTGR parameters.

The primary coolant system consists of the core and four to six primary coolant loops, each with its own circulator and steam generator. Helium gas, at a pressure of 700 psi (5 MPa), is pumped downward through the core and exits with a temperature of about 1370 °F (743 °C), considerably higher than for water-cooled reactors. The gas then passes into one of the pipes leading to a steam generator, where steam is raised for driving the turbogenerators. Above each steam generator is mounted a circulator which pumps the helium into the core.

The high reactor operating temperature is permitted by the gaseous form of the coolant and the good high-temperature characteristics of the core (there is no metal cladding that is sensitive to high temperature). This high temperature yields steam that can be converted to electrical energy with an efficiency of 39%, unusually high among thermal reactors. Moreover, the potential arises, with helium-driven turbogenerators, to improve even this high efficiency.

AUXILIARY SYSTEMS

The most noticeable auxiliary systems, shown in Figure 8-2, are the two or three auxiliary cooling loops. They are also contained in the PCRV and, in the event of failure of the main loops, can serve to remove the decay heat after reactor shutdown. However, the main cooling system is ordinarily the primary residual heat removal system following any shutdown.

Two identical systems are available for purifying the helium coolant. Each system uses filtration, adsorption, and a hydrogen getter to remove particulates and contaminant gases. One system operates while the other is shut down for decay and regeneration. The radioactive waste gas system is devoted largely to processing of gases released during regeneration of the purification system. These gases are separated into a radioactive component, which is ordinarily returned to the PCRV, and a stable component, which is released to the atmosphere. Liquid wastes arise only from decontamination operations, and the principal solid wastes are the tritium contaminated getters from the helium purification systems.

Figure 8-4. HTGR FUEL ELEMENT ARRANGEMENT.
HTGR fuel elements are arranged into stacks, which themselves are arranged in groups of seven; the central stack of each group has control rod channels. Note that the prestressed concrete vessel is lined with steel and protected with a thermal barrier. In addition, neutron reflector blocks surround the active core. (Figure courtesy of General Atomic Co.)

TABLE 8-1.

Representative Characteristics of High-Temperature Gas-Cooled Reactors

Core thermal power	2,900 MWth
Plant efficiency	39%
Plant electrical output	1,160 MWe
Core diameter	27.8 ft (8.5 m)
Core active height	20.8 ft (6.3 m)
Core power density	8.4 kW/liter
Number of core stacks (columns)	493
Number of fuel elements per column	8
Number of fuel elements	3,944
Element geometry	Hexagonal shape, 31 in high, 14 in across flats
Control rod type	Pairs of control rods in central stack of each refueling region (set of seven stacks)
Number of control rods	73 pairs
Reserve shutdown system	Spheres of boron carbide in carbon
Form of fuel	Fissile and fertile materials in different fuel particles, ^{235}U as UC_2, thorium+ bred ^{233}U in other particle type. Types have different coatings to facilitate separation.
Maximum fuel temperature	2,750 °F (1510 °C)
Average fuel temperature	1,450 °F (788 °C)
Average moderator temperature	1,320 °F (716 °C)
Coolant	Helium gas
Coolant flow rate	10.4×10^6 lb/hr (1.3 Mg/s)
Coolant pressure	700 psi (4.8 MPa)
Coolant temperature (inlet)	636 °F (336 °C)
Coolant temperature (outlet)	1,366 °F (741 °C)
Fuel exposure	98,000 MWd/Te
Fresh fuel assay (fissile particles)	93% ^{235}U (in initial loading)
Spent fuel assay (fissile particles)	30% ^{235}U (from initial loading)
Refueling sequence	One-fourth of the fuel per year
Weight of core and innards	6×10^6 lb (3×10^3 Mg)
Weight of PCRV (empty)	90×10^6 lb (4×10^4 Mg)

Source: General Atomic Co. specifications.

A steam generator isolation system is designed to prevent leakage of water or steam into the primary coolant. If the presence of water is detected, the defective cooling loop is isolated while the reactor is shut down, and the remaining loops provide cooling.

SAFETY SYSTEMS

The safety requirements of an HTGR are substantially different from those of water-cooled reactors. For one thing, the core provides a massive enough heat sink to lengthen by a large factor the time required for damage to occur to the fuel. Whereas decay heat can melt light-water fuel cladding within a minute or two of loss of cooling, HTGR fuel particles, with their ceramic coatings, can survive for as much as an hour. Moreover, the core's structural strength is provided by graphite,

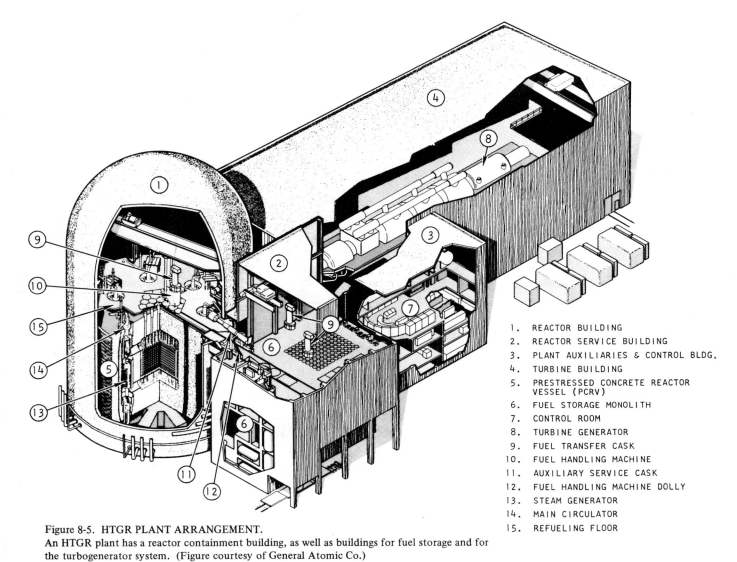

Figure 8-5. HTGR PLANT ARRANGEMENT.
An HTGR plant has a reactor containment building, as well as buildings for fuel storage and for
the turbogenerator system. (Figure courtesy of General Atomic Co.)

1. REACTOR BUILDING
2. REACTOR SERVICE BUILDING
3. PLANT AUXILIARIES & CONTROL BLDG.
4. TURBINE BUILDING
5. PRESTRESSED CONCRETE REACTOR
   VESSEL (PCRV)
6. FUEL STORAGE MONOLITH
7. CONTROL ROOM
8. TURBINE GENERATOR
9. FUEL TRANSFER CASK
10. FUEL HANDLING MACHINE
11. AUXILIARY SERVICE CASK
12. FUEL HANDLING MACHINE DOLLY
13. STEAM GENERATOR
14. MAIN CIRCULATOR
15. REFUELING FLOOR

whose strength increases as the temperature rises. On the other hand, the helium
coolant does not provide much cooling capacity unless it continues to be pumped
at high pressure. To make a complete loss of coolant extremely improbable, flow
restrictors are incorporated around PCRV penetrations to reduce helium loss should
the vessel integrity be violated there. As a result, helium is always presumed to be in
the system. Should all the primary cooling loops become unavailable (this is unlikely since they are largely independent), the auxiliary cooling loops can be activated and are sized to handle the decay heat. It is also worth noting that since the
coolant, helium, can be only in one phase and is nonreactive, certain complications
that may arise during accidents involving water coolant are eliminated.

HTGRs include a secondary containment structure (see Figure 8-5) as in other reactor plant types. The containment isolation and radioactive cleanup systems are similar to those of PWRs. In principle, though, it might be possible for less expensive containment systems to be used, considering the integrally contained nature of the primary coolant system in an HTGR.

NEUTRONICS, FUEL UTILIZATION, AND REACTOR OPERATION

Gas-cooled, carbon-moderated reactors have basic physics characteristics that are substantially different from those of water-cooled reactors. Use of carbon as a moderator implies that fission neutrons have to travel a much larger distance to reach thermal energies. The fuel distribution of an HTGR implies that the neutron energy distribution within the fuel pellets is not greatly different than it is in the moderator. As a result, HTGR fuel is subjected to more neutrons of intermediate energy than LWR fuel, and this can lead to a greater absorption of neutrons by fertile material, in this case ^{232}Th. This fact can be used to design a reactor with a relatively high conversion ratio.

However, the HTGRs offered commercially in the early 1970s generally had conversion ratios slightly less than 0.7, only slightly higher than that of LWRs. But the uranium utilization was also improved because the HTGR has a higher thermal efficiency (39%) than LWRs (33%). These factors led to a lifetime uranium requirement, assuming uranium recycle, of about 3000 tons of U_3O_8 as compared with more than 4000 tons for LWRs (with recycle). Gas-cooled reactors, including the HTGR, can be designed with significantly higher conversion ratios, as discussed in Chapter 10 (see especially Table 10-1) and Chapter 14 (Table 14-2).

The basic HTGR is designed with an average fuel burnup of 96,000 MWd/Te, about three times that of an LWR. This assumes replacement of a quarter of the fuel annually. Even though the thermal efficiency of an HTGR is high, the fuel burnup is higher than that of an LWR because the annual loading of fuel (both fertile and fissile) for an HTGR is about one-third the weight of that for an LWR. An HTGR designed for a higher conversion ratio typically includes a larger mass of thorium, and irradiates the fuel to a lower burnup (see Table 10-1).

Bibliography — Chapter Eight

"NTIS" indicates report is available from: National Technical Information Service, U.S. Department of Commerce, 5285 Port Royal Road, Springfield, VA 22161.

CONF-740501. "Gas-Cooled Reactors: HTGR's and GCFBR's," topical conference, Gatlinburg, May 7, 1974 (NTIS).
    Collection of technical papers on various aspects of gas-cooled reactors.
Dahlberg, R. C., Turner, R. F., and Goeddel, W. V., "HTGR Fuel and Fuel Cycle Summary Description," General Atomic Company report GA-A12801 (rev. January 1974).
    Describes HTGR core and fuel cycle.

EPRI NP-142. "Development Status and Operational Features of the High Temperature Gas-Cooled Reactor," Electric Power Research Institute report EPRI NP-142 (April 1976) (NTIS).

Summarizes developments in various nations pertaining to helium-cooled reactors, particularly thermal reactors.

ERDA-76-107. "Advanced Nuclear Reactors," U.S. ERDA report ERDA-76-107, (May 1976) (NTIS).

Very brief summary of advanced reactor systems.

Fleming, K. N., Houghten, W. J., and Joksimovic, V., "Status of the AIPA Risk Assessment Study for High Temperature Gas-Cooled Reactors," General Atomic Company report GA-A13970 (July 1976).

Summarizes progress in probabilistic study of HTGR risks.

WASH-1085. "An Evaluation of High-Temperature Gas-Cooled Reactors," U.S. AEC report WASH-1085 (December 1969) (NTIS).

An AEC assessment of the potential of HTGRs.

Wessman, G. L., and Moffette, T. R., "Safety Design Bases of the HTGR," *Nuclear Safety,* vol. 14, p. 618 (1973).

Summarizes features of the HTGR and considers these as they pertain to safe operation.

Part III
Uranium Resources, Advanced Fuel Cycles, and Nuclear Materials

Most nuclear facilities now operating are power plants utilizing reactors of the type described in Part II. These are supported by a relatively small number of enrichment and fuel fabrication facilities. However, any long-term substantial reliance on nuclear power will necessarily involve reactors of different types, as well as a more complete nuclear fuel cycle.

More advanced reactors will be needed because uranium resources are limited and, relatively speaking, the reactors described in Part II are profligate in their use of uranium. With nuclear power growth rates comparable to those that have been anticipated by federal agencies within the last decade, the plants built by the year 2000 could exhaust available resources. Continuation of nuclear power well into the next century will require a change either in the size of the nuclear power system anticipated or in the types of nuclear reactors and fuel cycles that are used.

Attention to the "back" end of the fuel cycle, particularly reprocessing and waste management, is needed because most advanced reactor systems depend on spent fuel reprocessing much more strongly than do light-water reactors and their contemporaries. Furthermore, two of the important public questions about nuclear power force attention toward the manner in which reprocessing and waste management are to be implemented. One question is how to control adequately the environmental emissions that may be associated with reprocessing and waste management operations, both now and in the future. The second is how to minimize the destabilization that could occur with the spread of materials that can be used to fabricate nuclear weapons. These questions, to a large extent, stand aside from the environmental and safety issues associated with the nuclear power plants themselves. They are, rather, connected with the disposition of nuclear materials in a global society that is not noted for its success at handling long-term environmental problems or at limiting the spread of nuclear weapons.

CHAPTER NINE

Uranium Resources and the Growth of Nuclear Power

THE EXTENT to which uranium resources can satisfy demands associated with current or future nuclear systems obviously depends on the form and availability of those resources. However, uranium resources are not well determined. As discussed below, most of the "resources" on which the future of nuclear power may depend are more or less speculative. Moreover, the effectiveness with which resources can be extracted depends on the economics and schedule of future growth of nuclear power. The discussion that follows is intended to convey an impression of the effective resource base.

Any uranium resource "crunch" would be caused by expansion of nuclear power in a way that made too great a demand on those resources. It is not possible, however, to predict accurately the manner in which nuclear power will grow. Still, a look at the manner in which it has grown, at the way predictions have changed, and at some of the factors that influence growth provides a framework in which to compare resources and demand.

In past analyses performed by utilities, the nuclear industry, and federal agencies of the United States, a comparison of uranium resources and nuclear power growth has usually led inescapably to the conclusion that an effective breeder reactor must be available in the near future. Such a reactor would produce substantially more fissile material than it destroyed, thereby making available excess fissile material for a growing nuclear economy. However, the realities of the late twentieth century may be so different from past predictions that alternative nuclear systems will play an important role.

The uncertainties and possibilities in the resource and demand question are treated in the next section. It is useful to begin by noting the contrast that historically led to the emphasis on breeder reactors. The Atomic Energy Commission, its successor the Energy Research and Development Administration, and presumably *its* successor, the Department of Energy (DOE), have for years cited economically useful U.S. uranium resources as about 3 to 4 million tons (1 ton = 0.9 Mg) of "yellowcake," U_3O_8. Considering that 1000 MWe reactors require a lifetime commitment of about 5000 tons of U_3O_8 (plus or minus 25%, depending on recycle, percentage tails, etc.), these resources could be expected to supply about 700 reactors. Since in many quarters a rapidly growing nuclear economy has been expected, surpassing 1000 reactors around the year 2000, an obvious conflict exists. This has led to emphasis on the liquid metal fast-breeder program. However, the realization has grown that enough uncertainties exist about our resource base, about the rate of growth (and ultimate size) of the nuclear power system, and about the alternatives available, to make this view of the problem and its solution simplistic. Such a view also ignores "externalities," such as the effect a choice can have on the proliferation of nuclear weapons.

RESERVES AND RESOURCES

Federal estimates of U.S. uranium resources have recently centered on 3.5 million tons of U_3O_8 at less than $30 per pound (0.45 kg). These resources have been categorized by state of discovery and by "forward cost," the estimated actual cost to mine the uranium and produce yellowcake, the U_3O_8 concentrate. Table 9-1 displays the amount of U_3O_8 that ERDA assigned to each category as of the beginning of 1977. The $30 to $50 category was added in 1977 and is the most uncertain. The discovery categories require some explanation. Only a small portion of the total are "reserves," i.e., developed or fully discovered resources. The remainder are "potential" resources and have been described by ERDA in the following way:

For its study of resources, ERDA subdivided potential resources into three categories: probable, possible, and speculative, to reflect the varying nature of the estimates depending upon the specific situation. Probable resources are those contained within favorable trends largely delineated by drilling data within known productive uranium districts. In this situation, favorable geologic characteristics of a formation are known from drilling or outcrop data, and quantitative estimates of potential resources are made by considering the size of the favorable areas and by comparing the geologic characteristics with those present in the areas with ore deposits.

Possible potential resources include those situations that are outside of identified mineral trends but which are in formations and geologic provinces that have been productive. Speculative resources are those estimated to occur in formations or geologic provinces which have not been productive but which, based on the evaluation of available geologic data, are considered to be favorable for the occurrence of uranium deposits. There is inherent uncertainty in these estimates, much more so for the speculative than the probable potential.

Many observers (Lieberman, for example) question the accuracy of the ERDA estimates. Considering the manner in which resource amounts in the more speculative categories must be determined, it would not be surprising if the estimates were in error.

The situation is complicated, not only by extremely scanty knowledge of uranium deposits, but by the rather hypothetical cost categories. The "forward

TABLE 9-1.
United States Uranium Resource Estimates[a]

Cost Category ($/lb U_3O_8)	Reserves	Potential			Total
		Probable	Possible	Speculative	
$10 or less	250,000	275,000	115,000	100,000	740,000
$10 – $15 increment	160,000	310,000	375,000	90,000	935,000
$15 or less	410,000	585,000	490,000	190,000	1,675,000
$15 – $30 increment	270,000	505,000	630,000	290,000	1,695,000
$30 or less	680,000	1,090,000	1,120,000	480,000	3,370,000
$30 – $50 increment	160,000	280,000	300,000	60,000	800,000
$50 or less	840,000	1,370,000	1,420,000	540,000	4,170,000

a. ERDA estimates for uranium resources as of January 1, 1977 (see text for discussion of other estimates). In addition, uranium that could be produced as a by-product of phosphate and copper production during the 1975 to 2000 period is estimated at 140,000 tons U_3O_8.

cost" is intended to reflect production costs. However, as extraction and processing costs change, a given deposit can change from one cost category to another. There is only a rough correspondence between ore grade and forward cost, and this correspondence changes as costs rise.

These cost changes are not to be confused with increases in the selling price of U_3O_8. For several years, the price of yellowcake was stabilized at around $8 per pound; but in the last two years or so, it has risen to about $40 per pound. These are changes, not in production cost, but in market price, reflecting uncertainties in future availability of uranium and strongly influenced by marketing decisions of uranium producers and their clients.

Uranium may be found in various geologic formations. Although large amounts of uranium are present in low concentrations in shales, phosphate deposits, granite, and seawater, the deposits that are commercially viable (less than $30 per pound forward cost) are relatively high-grade, with uranium oxide constituting more than 0.05% or 500 parts per million (ppm) of the ore. In the United States, significant amounts of ores have been found up to about 5000 ppm. This range (500 to 5000 ppm) comprises the bulk of the resources normally considered.

Most of these U.S. resources are present in sandstone deposits in several Western producing regions, notably the Colorado plateau and Wyoming basins. Wyoming, New Mexico, and Texas have recently been attracting most of the drilling activities, seeking new deposits. The geographical distribution of ERDA's "potential" resources are shown in Figure 9-1. (This is a 1976 figure; the totals for each category differ slightly from those given in Table 9-1.)

Although they cannot compete with high-grade ores, the low-grade uranium deposits offer some reserve in case the high-grade deposits are exhausted. The most nearly economical are the Tennessee shales, with uranium concentrations of about 10 to 100 ppm. The better of these shales (roughly 25 ppm and up) represent a very large resource, perhaps 10 to 15 million tons of U_3O_8. The economic cost of

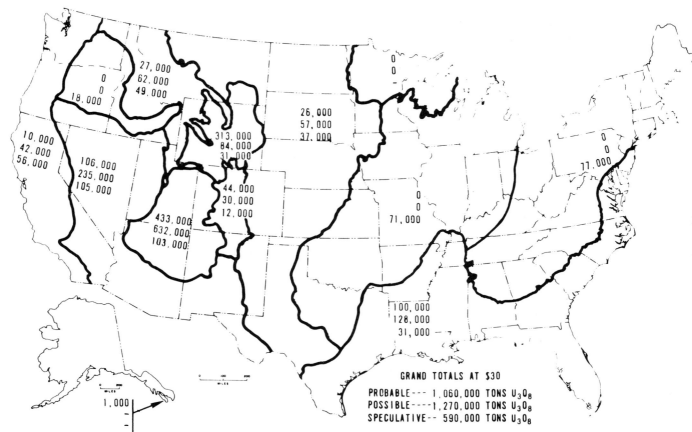

Figure 9-1. UNITED STATES POTENTIAL URANIUM RESOURCES.
Potential resources in the United States have been categorized as "probable," "possible," and
"speculative." The map shows estimates of these resources by region for uranium that is estimated to have forward costs of no more than $30 per pound of U_3O_8. These results are from
the 1976 interim report of the National Uranium Resource Evaluation. (Figure courtesy of
U.S. DOE.)

extraction is relatively high ($100 or more per pound), but might be tolerated,
particularly if it were regarded as a backup to more attractive, but more limited,
resources. However, such low-grade resources have little effective energy content
(i.e., the amount of electrical energy ultimately made available per mass of mined
ore); below about 50 ppm concentration, the energy extracted is no greater than
that from coal, assuming the uranium is used in the current LWR fuel cycle.
Considering how vigorous the processing of this ore must be (using sulfuric acid
leaching) and the problem of disposing of the tailings, these ores are very unattrac-
tive. The same may be said of granite ores with uranium contents about 10 ppm. Of
the low-grade sources, certainly the most attractive is seawater; but its concentration is so low (0.003 ppm) that economic extraction is not anticipated.

The geology of uranium deposits is a subject of high interest in connection
with resource assessment. We should point out that little uranium has been found in
the United States at intermediate concentrations, say 100 to 500 ppm, although
such deposits have been found in other countries. Generally, deposits are rather

localized, and this appears to be connected with the paucity of intermediate grade ores in the United States. This leads to some question whether experience in extraction of other resources is applicable to uranium. There is now renewed interest in searching for intermediate grade deposits.

In the last few years, observers have attempted to analyze the ERDA estimates or to make their own estimates based on data from the uranium extraction industry. No consensus has resulted from these analyses. Examination of the ratio of uranium finds to drilling efforts clearly indicates a drop in this ratio. These data have been interpreted to indicate that ERDA has substantially overestimated uranium resources, by a factor of two or three. On the other hand, it is also suggested that exploration activities have been rather limited in scope, and more complete exploration will lead to larger source estimates. In either case, it is agreed that more complete data are urgently needed as a basis for planning for nuclear power. Both ERDA (in its National Uranium Resource Evaluation) and the U.S. Geological Survey have made useful efforts in this direction. It appears that DOE will continue these efforts, perhaps changing the AEC classification system.

Another aspect of the resource question that must be considered is that, even for a fixed distribution of uranium resources, the time schedule for extraction may have a noticeable effect on the portion of the resource that the industry can actually extract. An intensive development program over a short period of time could extract more of a given resource base than a program that is stretched over a longer period. A working group of the National Academy of Science's study on nuclear and alternative energy systems suggested that only about two-thirds of the available uranium resource could actually be extracted. Even this would require a vigorous effort to extract the uranium as fast as possible. With 1.8 million tons of U_3O_8 as its number for likely resources at less than \$30/lb., the Academy group gave about 1.2 million tons as the portion that could be extracted, a number that is amusingly close to some relatively pessimistic estimates derived from drilling and finding data and considerably less than the 3.5 million tons estimated by ERDA and considered reasonable by a Ford Foundation study on nuclear power. It appears that 1 million tons is a safe lower limit on extractable uranium, but that substantially larger amounts may be discovered. Strategies for nuclear power should consider usable uranium resource bases as small as 1 to 1.5 million tons of U_3O_8.

As discussed in the next chapter, thorium deposits constitute a second nuclear resource. Although the extent of thorium resources is even less well known than that of uranium, it is widely believed that these two resource bases are comparable, as measured in mass. However, thorium is only a fertile material, that from which fissile material is generated. Nuclear systems that depend heavily on this conversion process will make much more effective use of thorium resources than light-water reactors make of uranium resources. This is true because LWRs convert little ^{238}U, which comprises 99% of natural uranium. As a result, it is improbable that thorium resources will be severely pressed in the next couple of centuries.

This examination of resources has concentrated on those in the United States. Although foreign resources are not better known than these, it is thought that United States and foreign resources are comparable. Strictly speaking, they cannot be considered independently because uranium is an international commodity.

NUCLEAR GROWTH AND URANIUM DEMAND

Predictions of the rate at which nuclear power will grow have undergone a great change in recent years. Once it had been established that nuclear plants were effective suppliers of electrical energy, there developed a greater and greater tendency to rely on them for future supply. In addition, the overall reliance on electrical energy increased significantly faster than growth of other energy types. In the decades before 1973, overall energy demand in the United States grew at a rate of about 4% per year, but electrical demand grew at 7% per year, which compounds to a doubling every decade. As a result, it was anticipated that electricity would come to dominate the energy supply by about 2000 and that nuclear power plants would constitute the bulk of the generating capacity. To put this in perspective, total generating capacity reached about 500 gigawatts (500,000 MWe) by 1976, of which about 10% was nuclear.[1] Many sources predicted that the nuclear generating capacity would exceed 1000 gigawatts by the year 2000. This would be 1000 plants of 1000 MWe size.

Few who are familiar with energy supply and demand would now predict continued growth on such a spectacular scale. Reasons for this altered perspective vary, as discussed in the next section, but it seems clear that overall energy use, electrical demand, and nuclear capacity will grow much more slowly than had been anticipated. The change is numerically large. During the first half of this decade, the Atomic Energy Commission predicted somewhat more than 1000 1000-MWe plants. Toward the end of this period, the AEC admitted the possibility of somewhat fewer than 1000. This view is adequately demonstrated in the AEC document WASH-1139(74), dated February 1974. Several cases were discussed therein (see Figure 9-2), projecting from 850 to 1400 plants by the year 2000.

Such a large growth rate bodes well for the health of the nuclear industry, particularly for the suppliers of light-water reactors. However, as noted earlier, economic uranium resources could not possibly supply 1000 reactors, each requiring 4000 to 6000 tons of U_3O_8 for its lifetime operation, and, in addition, supply the plants built after the year 2000. The AEC and industry had, in fact, anticipated that 1 to 3 million tons of U_3O_8 would be necessary by the year 2000 for cumulative requirements alone, including only initial cores and refueling up to the end of the century. This conflict between supply and demand led to a clear need for an effective breeder reactor, particularly if reliance on nuclear power was to increase in the next century.

Even with the introduction of a commercial breeder in 1993, now a very unlikely prospect, meeting the demand for uranium would be difficult. Figure 9-3 shows cumulative demand and lifetime commitments for the "low" Case A of WASH-1139(74), with 850 gigawatts by 2000. Cumulative demands are shown both with and without the fast breeder reactor, assuming that no ordinary converter reactors are built after 2000. Even for this "low" case, the ERDA estimates of

---

[1] In some regions of the country, such as the Chicago area, the nuclear portion was higher, as much as 30% or more. Note, however, that these figures represent generating *capacities*. The portion of electrical energy actually *generated* by nuclear plants may be more or less, depending on how much of the base load they supply and on their capacity factors.

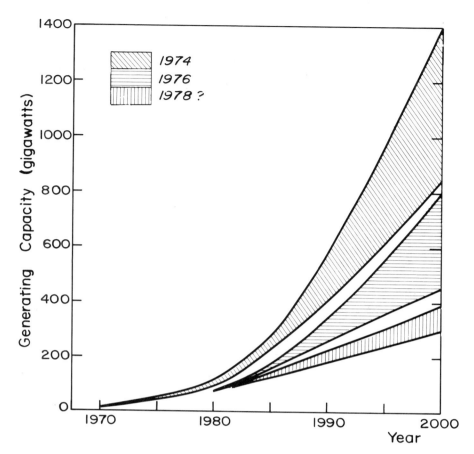

Figure 9-2. PREDICTIONS OF NUCLEAR ELECTRIC GENERATING CAPACITY.
Projections of nuclear power growth have changed rapidly in recent years. The range labeled
"1974" is from a 1974 report of the Atomic Energy Commission, WASH-1139(74). The
range labeled "1976" is from the Energy Research and Development Administration's national
plan for energy research, development, and demonstration, ERDA-76-1. The range labeled
"1978?" is intended to represent a more probable growth characteristic.

around 3.5 million tons would be exhausted. It would also be important for enrichment "tails" to be kept down to 0.2% ^{235}U.[2] If resources turn out to be significantly lower than 3.5 million tons, as certainly appears possible, a reduction in
growth rate would be necessary.

As is clear from Figure 9-3, with such a growth rate the fast breeder reactor
would have been useful, not to supply converter reactors with fuel, but to permit

---

[2] As described in Chapter 11 and Appendix F, an enrichment plant effectively separates the feed material into two components, one richer in the lighter isotope, and the other
partially depleted. The latter component is the "tails," which, for United States enrichment
plants, have contained from 0.2% to 0.3% ^{235}U, as opposed to the natural 0.7%. Operating at
0.2% rather than 0.3% significantly reduces the demand on resources. (It is important to
distinguish these values for the operating tails assay from the "transaction" assay, the nominal
value considered when the Department of Energy performs enrichment services. The two per-
centages may differ if uranium from federal stockpiles is added to the feed, rather than running
the enrichment plant at the transaction assay.)

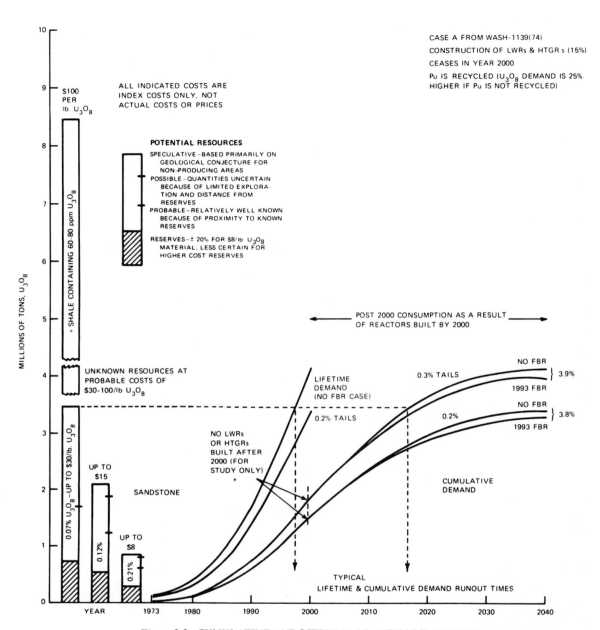

Figure 9-3. CUMULATIVE AND LIFETIME URANIUM DEMAND FOR A NUCLEAR SYSTEM WITH RAPID GROWTH.

The figure indicates the uranium resources used by a nuclear power system that grows to 850 gigawatts generating capacity by the year 2000. This is the "low growth" case of WASH-1139(74) (see Figure 9-2). The reactors built by the year 2000 are predominantly light-water reactors. Liquid metal fast breeder reactors are introduced during the 1990s, but only affect uranium needs significantly after 2000. Cumulative and lifetime demands of the pre-2000 system are little affected by introduction of the breeder, but the breeder is necessary to continue growth beyond 850 gigawatts. (Figure reproduced from ERDA-1.)

further expansion of nuclear power after the year 2000. Associated with even the "low" case cited above, a 1974 review of the liquid metal fast breeder reactor (LMFBR) program (ERDA-1) projected 1720 gigawatts nuclear by the year 2020, an unlikely prospect without a breeder.

This now appears to be a very unlikely prospect in any case, but was historically the prime reason for development of a fast breeder. In point of fact, an earlier introduction of the breeder, perhaps in the 1980s, had been anticipated; this would have had a more substantial impact on uranium requirements than Figure 9-3 indicates. However, it is still fair to say that the fast breeder was closely associated with expectations of a rapidly growing long-term reliance on nuclear power.

Expectations have changed radically since 1974. Although large uncertainties exist, it appears that the number of plants operating by the year 2000 will be less than half of that thought in 1974. Although government and industry groups still officially predict higher numbers, 300 to 400 gigawatts appears to many observers to represent the most probable range for nuclear capacity by the end of this century. Pushing the number substantially higher than 400 will require a large effort to build nuclear power plants. Holding the number much lower than 300 is unlikely in view of the fact that about half of this capacity is already operating or under construction.

All the commercial plants now being built are light-water reactors, with typical lifetime uranium requirements of about 4000 tons of U_3O_8, assuming the recycle of uranium and plutonium, and about 50% more without any recycle. Presuming that 1.2 million tons of U_3O_8 can be extracted, United States resources can support about 200 to 300 LWRs; more can be supplied only if resources are more extensive. We have not yet overcommitted ourselves, but some care must be taken that we do not soon do so.

A CHANGING PERSPECTIVE

The two fundamental factors that have caused a change in expectations for nuclear power are an expected lessening in the energy demand growth rate and an increasing tendency to turn to energy sources other than nuclear power. To some extent, these changes arise from strictly economic factors. But they also stem in part from general environmental and conservationist drives.

A slower rate of growth would, for example, result from higher energy prices, from energy conservation based on political or environmental concerns, and, perhaps, from ultimate saturation of the need for energy. The last factor will not affect energy demand in the next few years, but in a few decades it could begin to hold energy demand at a constant, or very slowly growing, level. The other two factors can, and certainly will, affect demand in the immediate future. The historical energy demand growth rate developed in an environment of constant, and even falling, energy prices. The doubling and even quadrupling of prices affects the economic balance between the use of energy and the institution of conservation measures, driving economic decisions in the direction of more efficient energy use and less growth in demand. Assuming institutional barriers are minimized and public awareness is maintained, substantially higher energy costs should result in substantially less demand in the long term. Political considerations, such as national energy self-sufficiency, and environmental considerations, such as possible pollution arising from power plants, will also tend to lessen growth, whether by governmental or individual choices. The net result will be a weakening of energy growth.

This would not necessarily imply a smaller number of nuclear power plants (unless some of the more absurd predictions are considered[3]). However, even relatively speaking, earlier predictions put an especially heavy emphasis on nuclear power. Decreased demand growth may be met with a more diversified mix of energy sources, rather than with whole-hearted reliance on nuclear power. If this occurs, whether for economic, political, or environmental reasons, projected nuclear power growth would be decreased even more drastically than projections of overall energy growth.

These considerations lead to the possibility that the growth of nuclear power will slow considerably after 1985. By that time, nuclear power plants already under construction will have been completed, and decisions made now can begin to have an effect. In some respects, the same is true of overall energy demand. Decisions taken now can have a greater effect on the last part of the century than on next year.

As suggested earlier, the United States nuclear generating capacity could easily end up in the 300 to 400 gigawatt range by the year 2000. What happens later is a matter of conjecture, but even this single datum indicates severely decreased pressure on uranium resources and a nuclear growth rate that admits many more possibilities than the LMFBR. The decreased pressure, in fact, reduces the need for a quick decision, permitting exploration of a broad range of possible nuclear systems and development of more information on energy demand and uranium resources. Whereas the fast breeder is appropriately associated with an economy based on rapid energy growth, it may be that other reactor types are consonant with a more settled, slowly growing energy economy.

[3] For example, with the high growth case of WASH-1139(74), which anticipated 1400 gigawatts nuclear in 2000, has been associated a projection of 4025 gigawatts in the year 2020 (see ERDA-1).

Bibliography — Chapter Nine

"NTIS" indicates report is available from: National Technical Information Service, U.S. Department of Commerce, 5285 Port Royal Road, Springfield, VA 22161.

The Committee on Nuclear and Alternative Energy Systems of the National Academy of Sciences — National Research Council conducted a study of various energy technologies from 1975 to 1977. Their results, including those on uranium resources and growth of nuclear power, should be published in 1978.

CONF-CF-74-5-26. "A Cursory Survey of Uranium Recovery for Chattanooga Shale," Oak Ridge National Laboratory report CONF-CF-74-5-26, (May 1974) (NTIS).

Cursory examination of how uranium might be obtained from shales.

EPRI EA-400. "Uranium Data," Electric Power Research Institute report EPRI EA-400 (June 1977) (NTIS).

Examines available estimates on uranium resources, as well as the methods by which the estimates are obtained.

EPRI EA-401. "Uranium Exploration Activities in the United States," Electric Power Research Institute report EPRI EA-401 (June 1977) (NTIS).

Examines current uranium exploration and considers how the resulting information might be improved.

ERDA-1. "Report of the Liquid Metal Fast Breeder Reactor Program Review Group," U.S. ERDA report ERDA-1 (January 1975) (NTIS).

On basis of growth rates stated in WASH-1139(74), estimates uranium requirements, leading to conclusion that LMFBR is required.

ERDA-1535. "Final Environmental Statement, Liquid Metal Fast Breeder Reactor Program," 3 vols., U.S. ERDA report ERDA-1535 (December 1975), with "Proposed Final Environmental Statement," 7 vols., U.S. AEC report WASH-1535 (December 1974) (NTIS).

Statement of environmental aspects of the LMFBR program, including radioactive emissions and resource utilization.

ERDA-76-1. "A National Plan for Energy Research, Development and Demonstration: Creating Energy Choices for the Future, 1976," vol. 1: "The Plan," U.S. ERDA report ERDA-76-1, U.S. Government Printing Office (April 1976).

Gives projections for the growth of nuclear power, as estimated in 1976.

Ford Foundation/MITRE Corporation, "Nuclear Power: Issues and Choices," (Ballinger, Cambridge, Mass., 1977).

Makes its own comparison of uranium resources and projected need, concluding LMFBR can be delayed.

GJBX-11(77). "Annual NURE Report 1976," Bendix Field Engineering Corporation report GJBX-11(77) (1977) (P.O. Box 1569, Grand Junction, CO 81501).

Brief report of progress on the National Uranium Resource Evaluation during 1976.

GJO-111(76). "National Uranium Resource Evaluation, Preliminary Report," U.S. ERDA report GJO-111(76) (1976) (available from Bendix Field Engineering Corporation, P.O. Box 1569, Grand Junction, CO 81501).

States estimated uranium resources as of January 1, 1976.

Lieberman, M. A., "United States Uranium Resources — An Analysis of Historical Data," *Science,* vol. 192, p. 431 (1976).

Examines the success rate for finding uranium, with the conclusion that ERDA uranium resource estimates are substantially too high. See also several letters on the subject in *Science,* vol. 196, p. 600 (1977).

Searl, M. F., "Uranium Resources to Meet Long Term Uranium Requirements," Electric Power Research Institute report EPRI SR-5 (November 1975) (NTIS).

After considering past coverage of uranium exploration activities, concludes that actual resources may be considerably greater than current estimates.

von Hippel, F., and Williams, R. H., "Energy Waste and Nuclear Power Growth," *Bulletin of the Atomic Scientists,* vol. 32, p. 14 (December 1976).

Argues that ERDA's projections of nuclear growth imply substantial inefficiencies in energy use.

WASH-1139(74). "Nuclear Power Growth, 1974-2000," U.S. AEC report WASH-1139(74) (1974) (NTIS).

Makes estimates of growth rate of nuclear power during the period 1974 to 2000.

CHAPTER TEN

Uranium Utilization in Advanced Reactors

THE REACTORS described in Part II were designed to produce electrical energy economically in a time when supplies of uranium were relatively assured and inexpensive. As exhaustion of uranium resources approaches and as extracted uranium becomes more expensive, concepts that are more frugal in their use of uranium become appropriate. With this end in view, reactors can be designed to husband available neutrons more carefully, so that a greater portion of them convert fertile material into fissile, thereby reducing, sometimes drastically, the depletion of fissile resources.

Effectively, the resources available to nuclear power are embodied in both uranium and thorium mineral deposits. Of these, only the 0.7% of uranium that is the mass 235 isotope is fissile, material that is readily split to release energy. The remaining 99.3% of the uranium and all the thorium constitute a potential resource that can be effectively realized only if it is converted to fissile material by the capture of neutrons. The main isotopes resulting from converting ^{238}U and ^{232}Th are ^{239}Pu and ^{233}U, respectively. Although the ordinary "converter" reactors described in Part II rely on such conversion to some extent, they do not emphasize this design aspect as much as they might. As a result, each requires about 3000 to 6000 tons of U_3O_8 to be committed to its operating lifetime, and at the end of its life, only a small portion of the fissile content of this commitment remains in the reactor.

Typically, most of the energy derived from these reactors arises from the fission of ^{235}U that is extracted from the ground. In the more neutron efficient of them, a comparable amount of energy is derived from fission of converted material. It is possible to design reactors which, proportionately speaking, convert more fertile material to fissile. A nuclear power system that incorporates such reactors as a basic component can ultimately derive most of its energy from fission of converted material. In this way, a greater portion of the energy potentially available from ^{238}U and ^{232}Th can be liberated.

144

These reactors fall into two classes: "thermal" reactors and "fast" reactors. The first class includes all the reactors of Part II, since they use a moderator to reduce neutron energies to the regime where fission cross-sections are high. Higher conversion ratios can be achieved by improving their neutronic design or by adopting basically different concepts. In many cases, such advanced thermal reactors can be designed to break even, i.e., to convert enough fertile material to fissile to maintain the fissile content. Going slightly beyond such a "near breeding" situation is possible, but such designs do not ordinarily hold much promise for producing a substantial excess of fissile material.

Producing such an excess is precisely the purpose of fast breeder reactors. These reactors are intended to produce enough fissile material to permit nuclear power to expand as rapidly as electrical demand. Conceived at a time when demand doubled every ten years, the fast breeder program has been intended to produce a system that doubles its fissile inventory each decade.

Chapters 13 and 14 describe the technical features of fast breeders and thermal "near breeders," respectively. Independently of these details, we can consider the characteristics of their fuel utilization as a basis for understanding the demand they make on uranium resources.

FAST BREEDER REACTORS

Fast reactors avoid thermalizing neutrons to take advantage of the fact that absorption by fissile materials of fast neutrons leads, on the average, to a larger number of fission neutrons. Numerically, the differences seem small, but it is the excess over 2 fission neutrons that is important. One neutron, on the average, is needed to continue the chain reaction; one must convert a fertile nucleus to one that is fissile (to replace the one that was destroyed in yielding energy and fission neutrons); and some excess must be available to overcome various neutron losses (to poisons, leakage, etc.). All told, significantly more than 2 neutrons must result from each neutron absorption by fissile material. Even for converter reactors, about 2 neutrons result (see Table 5-2 and 7-2), but a significant portion of these are lost to various absorbers. An effective breeder must overcome these losses.

All the three important fissile materials, ^{235}U, ^{239}Pu, and ^{233}U, have neutron yields per neutron absorbed (eta, η) that rise for absorbed neutron energies 1 MeV and above (see Appendix C, especially Table C-4 and Figure C-1). Plutonium 239 is most suitable for these purposes, because eta is in the vicinity of 3 over a substantial energy range. This is sufficiently in excess of 2 for breeding to be achieved. (See Figure 10-1.) The neutron yield from ^{233}U is not quite as high, but remains elevated over an even larger energy range, so that, in principle, it too could serve as the fissile load of a fast breeder reactor, even with a slightly moderated neutron spectrum. However, in the case of ^{235}U, eta is not large enough over a suitable energy range to permit the design of a practical breeder.

These considerations have led to the design of fast breeder reactors with plutonium as the fissile material and, ordinarily, with ^{238}U as the fertile loading. In all cases, the coolant and core structure are chosen to minimize moderation of the

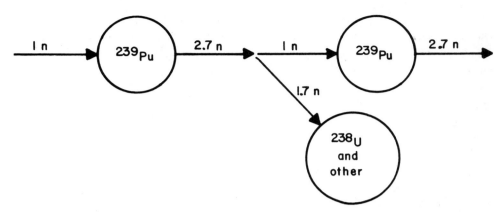

Figure 10-1. SCHEMATIC NEUTRON BALANCE FOR A URANIUM–PLUTONIUM FAST BREEDER REACTOR.
In a neutron spectrum typical of a fast breeder reactor, absorption of 1 neutron by ^{239}Pu yields, on the average, 2.7 neutrons. Of these, 1 is absorbed by ^{239}Pu, maintaining the chain reaction, and 1.7 are absorbed by other material, including fertile material, yielding about 1.2 fissile nuclei, predominantly ^{239}Pu. Thus more fissile material may be produced than is destroyed.

neutron energy, thereby achieving the highest possible yield of fission neutrons per fissile neutron absorption. Details of the predominant fast reactor types are given in Chapter 13.

The fuel utilization and production characteristics of breeder reactors can be characterized in a number of ways. The basic idea of a breeder is to increase conversion to the extent that the conversion ratio exceeds 1, so that excess fissile material is produced. To be more precise, the quantity usually associated with breeders is the "breeding ratio," the ratio of the number of fissile atoms produced to the number of the same kind that are consumed. This definition has historically been more precise than the "conversion ratio" (which does not distinguish between fissile types) and is appropriate for breeders where basically the same material is bred as is supplied. This has been the case anticipated for the liquid metal fast breeder reactor. For more complex situations, where more than one type of fissile material is involved, neither the conversion nor the breeding ratio is a complete characterization; detailed specification of input and output quantities is necessary. Such a situation could arise, for example, if the core fissile inventory were plutonium, but thorium was used to breed ^{233}U.

The breeding ratio actually achieved by a reactor will depend on how far the art of designing these reactors has advanced and particularly on advances in fuel design. It is expected, for the LMFBR, that early ratios will be 1.1 or slightly above, but that this will increase to at least 1.2 in later versions. On the other hand, a gas-cooled fast reactor has been designed to achieve a ratio of 1.4 (see Chapter 13 for more details).

The breeding or conversion ratio characterizes the reactor neutronically, but for the purposes of a breeder it is useful to indicate how it matches with the growth required for the nuclear power system. The important quantity in this context is the "doubling time," the period required for the reactor system to double the amount of fissile material that was initially committed. As usually defined, it is the time after which the excess productivity of one reactor is sufficient to fuel a second reactor. Since the electrical energy produced is roughly proportional to the total number of fissioned nuclei, it is easy to see how the doubling time is related to the breeding ratio. Defining the breeding gain as the excess of the breeding ratio over 1, the amount of excess fissile material produced per day for any given power level will be proportional to this gain, assuming other factors, such as the electrical energy per fission and the ratio of absorption to fission cross-sections, are fixed. The doubling time is simply the ratio of the reactor (or fuel cycle) inventory to this amount of excess material. Doubling time is therefore proportional to the fissile inventory divided by the breeding gain.

Strictly speaking, this definition would be correct only if the excess fissile material were held in storage until a sufficient amount accumulated to supply the entire fissile inventory of a new reactor. In the more likely situation, excess production of a number of reactors would contribute to a single new reactor, so that even the excess fissile material would soon be producing its own excess. In this situation, the rate of excess production has, effectively, to be compounded to yield a real doubling time.

In other respects, too, we have not defined doubling time precisely enough. For example, we have not given any attention to how differing input and output isotopic mixes affect the definition of doubling time. However, for present purposes, the most important clarification we can make is to distinguish the doubling time associated with only the fuel inventory of the reactor itself and that associated with the fuel, not only in the reactor, but also in the fuel cycle, i.e., in storage, reprocessing, fabrication, and so on. The practical doubling time of a breeder reactor system is the time it takes the reactor and its support facilities to yield enough excess fissile material to supply an additional reactor and *its* support facilities.

The doubling time is proportional to the fissile inventory. For a 1000-MWe LMFBR, the fuel would consist of 2 or 3 tons of fissile material, primarily plutonium, and about 50 tons of fertile material, e.g., ^{238}U. A comparable, perhaps somewhat smaller, amount of fissile material would be present in the fuel cycle undergoing processing, so that the total fissile inventory committed per reactor would be about 4 tons. A reactor with a smaller inventory would, of course, have a smaller doubling time, other things being equal.

Of course, in a situation where the nuclear power system is slowly growing, this may not be a primary consideration. Even then, however, the fissile inventory is of some importance because the initial loading of fast breeders must be supplied by thermal reactors. About 0.3 tons of plutonium could be recovered per year from the spent fuel of a typical uranium fueled light-water reactor (see Figure F-2), of which about 0.2 tons is fissile (^{239}Pu and ^{241}Pu). Roughly speaking, then, about 20 reactor-years of operation would provide the fissile inventory associated with a single fast breeder. If hundreds of LWRs were operating, several breeders could be started up per year. Decrease in the required inventory would obviously increase the rate at which breeders could be started up.

The other fuel that must be supplied to a breeder is fertile material, either ^{238}U or ^{232}Th. Depleted uranium is available as tails from enrichment plants. The current U.S. stockpile of more than 200,000 metric tons of depleted uranium would supply a large system of fast breeder reactors for about a century. Thus fast breeders could receive both their fissile and fertile requirements as by-products of the light-water reactor system. They could also draw on what appear to be adequate thorium resources. In this sense, fast breeder reactors do not utilize uranium resources. Construction of a fast breeder is intended to decrease dependence on natural fissile deposits.

HIGH CONVERSION THERMAL REACTORS

Advanced thermal reactors may be designed to utilize available neutrons much more carefully than do the converters described in Part II. Sufficiently high conversion ratios may be achieved so that, after an initial running period, little or no additional fissile material need be supplied from the resource base. Achievement of a conversion ratio of 1, so that for each fissile nucleus consumed another is supplied by conversion of fertile material, means that no external fissile supply is required. In fact, when a given reactor is decommissioned, its fissile inventory is adequate for operation of a new reactor replacing it. Effectively, operation of such a reactor entails no depletion of fissile resources.

It is possible to approach or slightly surpass this break-even state with thermal reactors. In most current thermal reactors, for each neutron absorbed by fissile material, the average number of neutrons produced is very close to 2. For break-even, one neutron must be allotted to continuation of the chain reaction and another must successfully convert a fertile nucleus to one that is fissile. Achieving break-even, therefore, requires raising the neutron production ratio enough above 2 to overcome various neutron losses in the reactor. This requires careful choice of fuel type and moderating conditions and careful minimization of neutron losses.

In current light-water reactors, a ratio of only about 0.6 is achieved for two fundamental reasons: the water moderator and the reactivity control systems absorb a significant portion of the neutrons from fission, and the nuclear fuel itself cannot practically sustain a conversion ratio of 1 because the ^{235}U–^{238}U–^{239}Pu combination is not good neutronically. By the latter we mean that ^{235}U and ^{239}Pu produce such a small excess of neutrons (per neutron absorbed) above the 2 required for break-even, that any small unavoidable losses will depress the conversion ratio below 1. In fact, the neutron production ratio (see eta, η, in Appendix C) for ^{239}Pu is actually less than 2 for a large range of energies that would be substantially populated as neutrons are thermalized, so that a spectrum-averaged value is typically 1.8 or 1.9, well below the absolute minimum for break-even. (Remember that this entire discussion refers to *thermal* reactors.)

Were ^{232}Th used as the fertile material, the situation would be different. First, ^{232}Th has a higher neutron absorption cross-section than does ^{238}U (Appendix C), so that more conversion takes place in the former. As a result, more ^{233}U is produced than ^{239}Pu in equivalent circumstances. (This means, of course, that

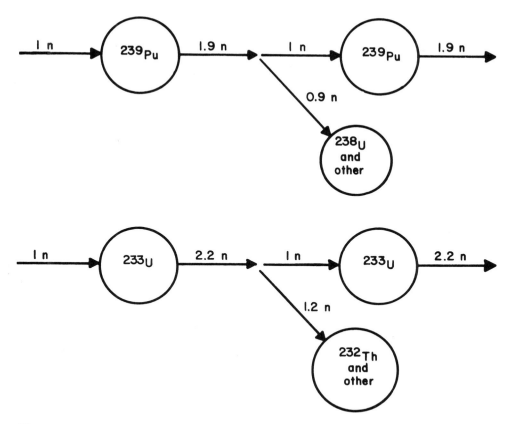

Figure 10-2. SCHEMATIC NEUTRON YIELD AND CONVERSION PROCESSES WITH ^{239}Pu AND ^{233}U IN A THERMAL REACTOR.

The upper portion of the figure shows the neutron yield resulting from absorption of 1 neutron by ^{239}Pu. Of the 1.9 neutrons resulting, only 0.9 is available for capture by ^{238}U and other nonfissile materials, not enough to replace the ^{239}Pu nucleus that is destroyed. On the other hand, with ^{233}U and ^{232}Th, the neutron yield of 2.2 is enough to continue the chain reaction and replace the destroyed ^{233}U, providing that 1 of the 1.2 excess neutrons converts a ^{232}Th nucleus.

more fissile material must be supplied initially to overcome the greater absorption of the ^{232}Th). Furthermore, ^{233}U is a much better thermal fuel than ^{239}Pu, and is even superior to ^{235}U. (This is the reverse of the situation in today's LWRs, where the produced fissile species, ^{239}Pu, is a poorer thermal fuel than the initial fissile type, ^{235}U.) The reason for ^{233}U's superiority, as depicted in Figure 10-2, is that, for each neutron absorbed by ^{233}U, noticeably more than two neutrons are produced. Eta (η) typically has a value of 2.2 or more in a thermal spectrum. The comparable number for ^{235}U is very close to 2, and we noted above that ^{239}Pu is even worse.

The best possible situation has ^{233}U and ^{232}Th as the fuel, with ^{232}Th being converted to replace the ^{233}U that is consumed. Since ^{233}U does not occur naturally (due to its "short" half-life), this condition must be reached by beginning with natural ^{235}U (or by-product ^{239}Pu) as the fissile fuel, in the presence of

TABLE 10-1.
Thermal Reactor Fuel Characteristics

Characteristics[a]	BWR	PWR	Std HTGR	High-Gain HTGR	CANDU (Thor)	CANDU	ATR
Thermal efficiency (percent)	34	33	39	39	30	30	31
Specific power (MWth/Te)	28	38	82	69	25.6	22	13
Initial core (and blanket)[b] average							
Irradiation level (MWd/Te)	17,000	22,600	54,500	33,500	18,500	6,900	13,700
Fresh fuel assay (Wt % 235U)	2.03	2.26	93.15	93.15	93.15	0.711	1.2
Spent fuel assay (Wt % 235U)	0.86	0.74	~60	~60	~60	0.31	0.33
Fissile Pu recovered (kg/Te)[c]	4.8	5.8	—	—	—	1.7	4.3
Feed required (ST U_3O_8/MWe)[d]							
at 0.2% tails	0.494	0.422	0.367	0.388	0.520	0.199	0.516
at 0.3% tails	0.581	0.498	0.456	0.419	0.646	0.199	0.577
Feed required (ST ThO_2)[b,d]	—	—	42.5	52.8	160	—	—
Separative work required (SWU/MWe)[d]							
at 0.2% tails	239	222	366	336	519	—	—
at 0.3% tails	185	174	311	286	441	—	—
Replacement loadings (annual rate at steady state; 75% capacity factor)							
Irradiation level (MWDth/Te)	27,500	32,600	95,000	48,800	27,000	9,600	20,000
Fresh fuel assay (Wt % 235U)	2.73	3.21	93.15	93.15	93.15	0.711	0.711
Spent fuel assay (Wt % 235U)	0.84	0.90	30	30	—	0.15	0.2
Fissile Pu recovered (kg/Te)[c]	5.9	7.0	—	—	—	2.3	—
Feed required (ST U_3O_8/MWe)[d]							
at 0.2% tails	0.144	0.154	0.085	0.045	0.020	0.125	—
at 0.3% tails	0.179	0.191	0.106	0.055	0.024	0.125	—
Feed required (ST ThO_2/MWe)[b,d]	—	—	8.7	19.7	41.5	—	—
Separative work required (SWU/MWe)							
at 0.2% tails	105	117	85	45	19	—	—
at 0.3% tails	84	94	73	38	16	—	—
Replacement loadings (annual rate with Pu recycle[e], 75% capacity factor)							
Fissile Pu recycled (kg/MWe)	0.163	0.167	—	—	—	—	0.3
Fissile Pu recovered (kg/Te)[c,f]	8.1	9.5	—	—	—	—	5.5
Fissile 233U recovered (kg/MTTh)[b]	—	—	—	—	—	—	—
Feed required (ST U_3O_8/MWe)[d,g]							
at 0.2% tails	0.121	0.129	—	—	—	—	—
at 0.3% tails	0.148	0.158	—	—	—	—	—
Separative work required (SWU/MWe)[d]							
at 0.2% tails	82	93	—	—	—	—	—
at 0.3% tails	66	75	—	—	—	—	—

TABLE 10-1.
Thermal Reactor Fuel Characteristics (continued)

Characteristics[a]	BWR	PWR	Std HTGR	High-Gain HTGR	CANDU (Thor)	CANDU	ATR
Lifetime[h] commitment required, 40-year life, ST U3O8/1000 MWe[b] 40 (Replacement requirement) + initial core and blanket							
Without Pu recycle							
at 0.2% tails	5,130	5,330	—	—	—	4,160[i]	—
at 0.3% tails	6,340	6,550	—	—	—	4,160[i]	—
With Pu recycle							
at 0.2% tails	4,020	4,100	—	—	—	—	2,600
at 0.3% tails	4,900	5,050	—	—	—	—	2,660
With thorium and 233U recycle							
at 0.2% tails	—	—	2,980	1,920	1,440	—	—
at 0.3% tails	—	—	3,700	2,400	1,780	—	—

a. MWth is thermal megawatts; MWe is net electrical megawatts; MWDth is thermal megawatt days; Te is metric tonnes (thousand of kilograms) of uranium; and ST U3O8 is short tons of U3O8 yellowcake from an ore processing mill. One SWU is equivalent to one kg of separative work.

b. Information requested not previously in Table 12, WASH 1139(74), page 24.

c. After losses.

d. For replacement loadings, the required feed and separative work are net, in that they allow for the use of uranium recovered from spent fuel. Allowance is made for fabrication and reprocessing losses.

e. Plutonium available for recycle ratchets up each pass because not all the plutonium charged is burned. Therefore, more plutonium is recovered from mixed oxide fuel than from standard uranium fuel, and this increment increases with each cycle (5–6 years per cycle) requiring several passes to reach steady state. The data shown represent conditions for the 1980s when most reactors will be discharging fuel which has only seen one recycle pass.

f. Average for all fuel discharged with full recycle of self-generated plutonium. For mixed oxide fuel (natural U spiked with self-generated plutonium) the spent fuel from BWRs contains 15.1 kg Pu per Te and from PWRs, 18.7.

g. Includes natural uranium to be spiked with plutonium; 0.0087 ST U3O8/MWe for BWR and 0.0067 for PWR.

h. Lifetime commitments assume operations at 40% capacity factor (CF) for the first year, 65% CF for the next two years, followed by 12 years at 75% CF. Therefore, CF drops two points per year, reaching 25% in the last (40th) year.

i. A burnup of 9600 MWd/Te has not yet been achieved, so that the current generation of CANDUs requires about 15% more uranium feed than indicated in this table.

Source: Adapted from "Report of the Liquid Metal Fast Breeder Reactor Program Review Group," U.S. ERDA report, ERDA-1 (January 1975).

which ^{233}U is produced. Since there is little doubt that reactors using ^{233}U–^{232}Th can be built to achieve a conversion ratio close to 1, what has to be supplied from external sources is the fissile material to produce enough ^{233}U. This must be either ^{235}U or ^{239}Pu.

Some inkling of the potential value of the thorium cycle can be obtained from Table 10-1, which gives the fuel needs of a variety of reactor types under a number of conditions. The results for the BWR and PWR correspond to currently constructed plants, as do those for the ordinary CANDU described in the last chapter. The quantity that we will concern ourselves with, for simplicity, is the lifetime uranium requirements. Note that the LWR results depend on whether fuel is reprocessed, in particular whether plutonium is recycled, and on the percentage ^{235}U left in the enrichment tails. Neither of these factors affects the standard CANDU because, as currently operated, reprocessing is presumed not to occur, and only natural uranium is used as the fuel. CANDU without reprocessing still compares well with the best that LWRs can do (see Table 10-1). This is due to the higher conversion ratio possible when heavy water is used as moderator-coolant and to the reduced losses to control systems that are made possible by on-line refueling.

The "standard" HTGR is the only reactor for which current designs depend on the thorium cycle. As discussed in Chapter 8, HTGR fuel particles initially contain ^{235}U and ^{232}Th. Eventually, noticeable concentrations of "bred" ^{233}U will reside in the thorium particles. After reprocessing, this ^{233}U is recycled back to the reactor, as a result of which the lifetime feed of natural uranium is less than 3000 tons of U_3O_8 as compared with the more than 4000 tons for the current water-cooled reactors. The lesser figure is also due to other factors, such as the use of carbon and helium as moderator and coolant, respectively, and the higher thermal efficiency of the HTGR.

However, better results can be obtained even for water-cooled reactors by introducing a thorium cycle. Preliminary studies indicate that in LWRs, a lower lifetime requirement can be achieved simply by replacing most of the ^{238}U in the present fuel loads by ^{232}Th. This initial use of highly enriched uranium together with ^{232}Th, followed by recycle of bred ^{233}U, makes this concept similar to the HTGR scheme just described. However, the fuel form would be very different, and the continued use of light water would inevitably cost neutrons (and hence conversion ratio). The net effect, though, is still a reduction by about 10% or 20% in uranium requirements.[1] However, this presumes no redesign of the core, and takes no account of altered cooling, control, and other characteristics. An effort to optimize the core neutronically and in these other respects would certainly cause greater improvement in the uranium requirement, as discussed in Chapter 14. The extreme case of such redesign is the light-water breeder reactor, which is intended to achieve conversion ratios slightly greater than 1. More moderate design alterations could presumably be incorporated with some success in present LWR power plants.

However, the CANDU and HTGR are ultimately more susceptible to improve-

---

[1] See Chapter 14 for further discussion.

ment because of their inherently better moderator systems (from the neutronic point of view). Table 10-1 shows results for a "high-gain" HTGR which has a modified fuel-moderator ratio and a more frequent refueling schedule to achieve better conversion ratio. The decrease in the U_3O_8 requirement is substantial, a reduction of one-third.

The "thorium CANDU," also shown, involves a greater change, i.e., from natural uranium to ^{233}U–^{232}Th fuel. Since Canada intends to build none of the enrichment plants that would be necessary to secure relatively pure ^{235}U, plutonium is necessary as an intermediary. Present CANDUs produce ^{239}Pu and, as in LWRs, it contributes to the heat produced by the reactor. Rather than leave this plutonium in the spent fuel, as is now done, this fuel would be reprocessed to obtain Pu, which would then be used as feed to an initially plutonium-thorium core. By adding enough plutonium, sufficient ^{233}U can be bred to fuel the reactor at a conversion ratio of 1, so that the reactor is self-sustaining thereafter. A figure of 1200 Mg (1320 tons) of U_3O_8 is named by AECL as the total feed to the reactor system to eventually produce the ^{233}U for one self-sustaining reactor (and the number may even be less if the plutonium stage could be avoided by beginning with highly enriched uranium). This last condition is much desired; the 1320 tons not only supplies the reactor for its lifetime, but also the reactor replacing it at its decommissioning. (Table 10-1, from the LMFBR review, attributes a somewhat larger requirement to the thorium CANDU.)

Improving the moderator is also a possibility considered for U.S. water-cooled reactors. This could be done by replacing part or all of the light-water coolant by heavy water. This has the direct effect of reducing the loss of neutrons to moderator absorption. It also has the more subtle effect, for a given moderator to fuel ratio, of less effectively moderating neutrons. At first blush, this may seem an unfortunate result. However, there is evidence that under some conditions, particularly with a thorium-uranium fuel cycle, this "hardening" of the neutron spectrum may increase the conversion ratio. (See discussion of spectral shift control in Chapter 14.)

Finally, the molten salt breeder reactor, also described in Chapter 14, appears to have potentially as high a conversion ratio as any thermal reactors. It would, in fact, be a thermal breeder. However, it is not at present being developed.

Reactors with conversion ratios approaching 1 appear to have a feature that is quite unfortunate — the fuel must be reprocessed quite often. The distinction between refueling frequency and reprocessing frequency is important. The first refers to how often the reactor is refueled, without regard to how much of the core is replaced. The reprocessing frequency is associated with fuel lifetime, i.e., with how often a given fuel assembly is replaced and reprocessed. In the CANDU, for example, which has on-line refueling machines, refueling is essentially continuous. Conversion ratios between 0.8 and 0.9 can be achieved with thorium fueling even with relatively high burnup (long-lived) fuels, irradiated to as much as 30,000 MWd/Te. However, to achieve a ratio of 1, the fuel can be used for only a short time (with about 10,000 MWd/Te burnup) before refueling. This reduces the portion of neutrons lost to absorption by various poisons. The cost, though, is that much more reprocessing and fabrication are required, with a direct economic penalty (see Chapter 11). This is also true of other high conversion systems. The one system that would not be disturbed by this requirement is the molten salt breeder reactor, which has on-line reprocessing.

Although we have spoken exclusively of lifetime uranium requirements, this becomes a less useful indicator as the conversion ratio approaches 1. For the commercial power plants of Part II, only a small portion of the lifetime requirement is needed for the initial core, so that this requirement truly represents a depletion of resources. For systems with higher conversion ratios, much of the lifetime requirement is used to start the reactor or to reach equilibrium. This portion of the lifetime requirement is *a commitment that is not lost.* It could serve for generation after generation of reactors. The portion of the lifetime requirement that is lost is the annual make-up requirement. This is, of course, zero for a conversion ratio of 1. To make this distinction clear, Table 10-2 shows the commitment to initial loading (or to reaching equilibrium) and to annual refueling for both commercial and advanced reactors. For the "advanced" reactors, the annual refueling requirement is very small, so that little resource depletion occurs.

TABLE 10-2.

Representative Uranium Requirements for 1000-MWe Thermal Reactors[a]
(tons of U_3O_8 at 0.2% tails)

	Initial Loading (Commitment for Achieving Equilibrium)	Annual Refueling at 75% Capacity Factor	30-Year Lifetime Requirement for Initial Reactor[b]
Commercial Reactors			
LWR	452	125 (200)[c]	4,080 (6,400)[c]
CANDU[d]	199	150[d] (74)[e]	4,910[d] (2,420)[e]
HTGR	367	85 (145)[c]	2,980 (4,700)[c]
Advanced Reactors[f]			
Thorium PWR	(800)	85	3,350
High-gain CANDU	520(850)	20	1,440
High-gain HTGR	338(570)	45	1,920
Break-even CANDU	(1,320)	0	1,320
Break-even LWBR	(2,000)	0	2,000

a. Many of the entries in the table are from Table 10-1. Others are adapted from other sources (see bibliography at end of chapter). The most significant difference with Table 10-1 is the CANDU annual refueling requirement, which is based on burnup at the newer plants (8200 MWd/Te), rather than the 9600 MWd/Te assumed in 10-1. In any case, the numbers are intended to be representative, not definitive. Unless otherwise specified, they assume recycle of both uranium and plutonium.

b. For the purpose of this discussion "lifetime requirement" is that for 30 years at 75% capacity, which is roughly 22.5 full-power years, or 40 years at 56% (the average capacity factor used in Table 10-1).

c. This requirement assumes no recycle.

d. The ordinary CANDU is the only system in the table that does not usually assume recycle of bred fissile. As noted, the annual refueling requirement would be halved if plutonium were recycled. See also note a.

e. This requirement assumes plutonium recycle.

f. The advanced reactors all require recycle. For the break-even reactors, the "commitment for achieving equilibrium" includes that for developing the entire fissile inventory for a reactor, including the portion in process in the fuel cycle.

For break-even thermal reactors, it is useful to emphasize the distinction between the equilibrium state, when no demand is made on uranium resources, and the "prebreeder" stage, when considerable demand is made. During prebreeding, an inventory of fissile material (^{233}U) is developed, which is sufficient to sustain the reactor and its fuel cycle. Development of this inventory requires a substantial utilization of ^{235}U or ^{239}Pu.

URANIUM DEMAND WITH ADVANCED REACTORS

Since uranium resources are limited, it is useful to ask what the total demand on these resources will be. Figure 9-3 presented this information for a rapidly growing nuclear power system that included LWRs and, after 1993, LMFBRs. It presented both cumulative demand and lifetime demand for the entire system. It is now apparent that nuclear capacity will not grow this rapidly, as was suggested in Figure 9-2.

Even in a low-growth system, the demand on uranium will depend strongly on the type of reactors that are built after 1985 and the time when these reactors are introduced. A careful distinction, too, needs to be made between commitments that are recoverable and those that are not. Makeup at annual refueling is lost, in contrast to initial loading, which will serve as well in a replacement reactor. A final complication is caused by a nuclear power system that involves two or more reactor types, with fissile transfers from one type to another.

Liquid metal fast breeder reactors are a case in point. They operate on uranium and plutonium from the LWR system, so that they do not directly consume resources. In fact, their purpose is to decrease resource use. As seen in Figure 9-3, they would have no noticeable effect on resource depletion in this century, but would provide a limit to ultimate requirements that is essentially independent of the rate of growth of nuclear power in the next century.

If, however, the growth rate is low, then excess fissile production would be available for other uses. In particular, if this excess is used in thermal reactors, full use can still be made of breeder capabilities. If this were anticipated, however, it would be advantageous to introduce thorium as part of the breeder's fertile material, thereby producing ^{233}U for use in thermal reactors. As noted earlier, this isotope is a superior fissile feed at thermal energies.

On the other hand, if the nuclear power growth rate is slow, high-conversion thermal reactors can be attractive. It is useful to display the potential effects of such reactors along the lines that were used in the last chapter (Figure 9-3). To begin with, we take the most conservative and simple point of view by estimating the total commitment of uranium for the entire reactor system, using the lifetime requirements associated with each reactor type. We perform a calculation for two possible growth patterns, one with 250 gigawatts nuclear by the year 2000, and the other with 500. These effectively bracket what appears to be the most probable result. Both systems will have large contingents of ordinary LWRs since so many have already been ordered and, moreover, new designs take time to develop. However, after 1990, new reactors can have increasingly high conversion ratios. Reactors

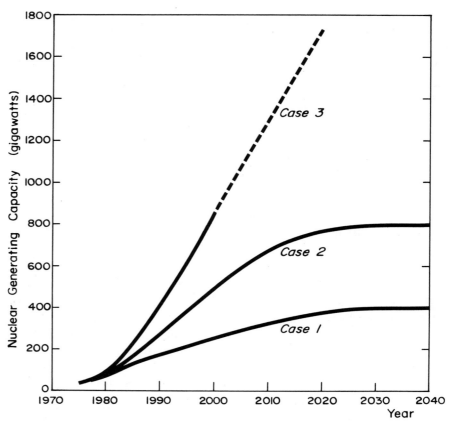

Figure 10-3. LONG-TERM NUCLEAR POWER GROWTH CURVES.
The figure presents three possibilities for the long-term growth of nuclear power. Case 1 is a low-growth case, where the nuclear electrical generating capacity expands to 250 gigawatts by the year 2000, then grows more slowly to a maximum capacity of 400 gigawatts. Case 2 indicates high growth: 500 gigawatts by 2000 and a maximum capacity of 800 gigawatts. For comparison, the old "low-growth" case of WASH-1139(74) (see Figures 9-2 and 9-3), is given in case 3. Uranium requirements corresponding to these cases are given in Figure 10-4.

built after 2000 could be self-sustaining (conversion ratio 1). To illustrate the long-term impact of such reactors, we will presume each of these two cases to evolve into a system that has a permanent generating capacity somewhat larger than the capacity of the year 2000. The 250-gigawatt system would eventually grow to 400 gigawatts, the 500 to 800. This later growth would presumably proceed more slowly, leveling out to the final figure by about 2030, but the detailed pattern will have little effect on the total resource commitment. As pre–2000 reactors are decommissioned, it is assumed that they are replaced with "break-even" reactors.

The two possible growth patterns are shown in Figure 10-3 as cases 1 and 2. (Case 3, the "low-growth" pattern of WASH-1139[74], is shown for comparison.) To compare possible uranium lifetime commitments, we postulate that all reactors completed by 1990 are conventional light-water reactors. The bulk of the reactors built by that date will be ones that are already on order. To each conventional LWR, we assign a lifetime uranium commitment of 4000 tons of U_3O_8. In fact, if it were desired, it seems clear that retrofits could decrease this requirement. Post-1990 reactors would have increasing high-conversion ratios and, conversely, decreasing annual refueling requirements. For present purposes, we assume power plants have the following requirements (in tons of U_3O_8):

Completion Date	Lifetime Commitment	Annual Refueling (75% capacity)
Up to 1990	4000	125
1990 − 1995	3000	85
1995 − 2000	2500	60
Post 2000[a]	1500	0

a. The "post-2000" break-even reactor is assumed to require 1500 tons of U_3O_8 to supply the fissile for the reactor's core plus fuel cycle inventory. This is somewhat more than the amount claimed for a thorium-uranium CANDU. Even if the requirement is raised slightly, this has little effect on the results that follow.

The lifetime uranium commitments that correspond to the growth patterns of Figure 10-3 are shown in Figure 10-4. Cases 1 and 2 have a curious shape in the period after 2000 because of the replacement of decommissioned low-conversion reactors by new self-sustaining systems. For Cases 1 and 2, the commitment shown is adequate for permanent systems of 400 and 800 gigawatts, respectively. For Case 3, continuation of the system at the 850-gigawatt level would require additional commitments after the year 2000 or the introduction of fast breeder reactors.

Essentially all the total commitment of Case 3 (about 3.4 million tons of U_3O_8), and about 60% of the smaller commitments of Cases 2 and 3 represent actual depletion of fissile resources, associated with annual refueling requirements. Most of this depletion arises from the use of low-conversion ("4000 ton") LWRs. Building a 400-gigawatt self-sustaining system from scratch would, for example, have required only 0.6 million tons of U_3O_8, as compared with the total 1.4-million-ton commitment shown on Figure 10-4. This portion of the commitment is associated with the permanent inventory required by 400 break-even reactors. Similarly, 1.2 million tons are the U_3O_8 feed requirement to make up the permanent inventory of the 800-gigawatt system. If the uranium feed per reactor were 2000 tons (instead of the 1500 tons postulated), the total requirement for the 800-reactor system of Case 2 would be 3.1 million tons, of which 1.6 million would have supplied the system's permanent inventory.

The point of Figure 10-4 is to show the manner in which high-conversion thermal reactors can constitute a long-term nuclear generating capacity with only a limited contribution from uranium resources. The bulk of the system's energy is effectively drawn from thorium resources. These resources are, of course, not unlimited either. But it appears that they will be ample, especially if thorium is recycled after an adequate storage time. (Immediate recycle would be difficult because of the [228]Th contaminating the [232]Th; see next chapter.)

Alternatively, after the initial period of light-water reactors, fast breeders

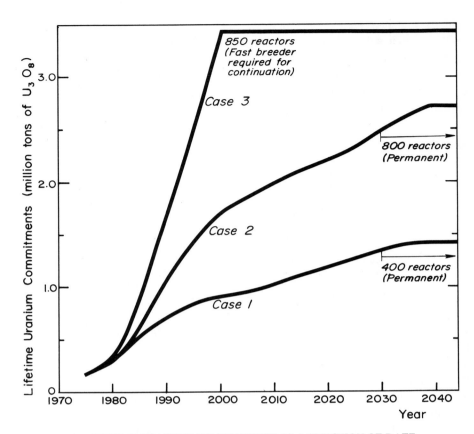

Figure 10-4. LIFETIME URANIUM COMMITMENTS AS A FUNCTION OF DATE.
For the three cases given in Figure 10-3, this figure presents lifetime uranium commitments for
reactors built by a specified date. For cases 1 and 2, the date when specified permanent
nuclear capacities are reached is shown; these cases presume that, as the technology is developed, thermal reactors with higher and higher conversion ratios are built until, by the year
2000, only "break-even" reactors are being constructed (see text). For Case 3, continuation of
the nuclear capacity at 850 gigawatts would require introduction of a fast breeder reactor.

could supply the bulk of nuclear generating capacity. They would, in fact, be
necessary for a capacity that grows rapidly for a long period of time. For a more
slowly growing system, some mixture of fast breeders and thermal converters, or an
exclusively high-conversion thermal system, would be possible. In both cases, an
important requirement for efficient uranium utilization would be expected to be
the availability of fuel reprocessing. However, in the event that nuclear growth is
slow and available resources are high, it would be feasible to rely on improved once-
through systems for a substantial period.

Bibliography — Chapter Ten

"NTIS" indicates report is available from: National Technical Information Service, U.S. Department of Commerce, 5285 Port Royal Road, Springfield, VA 22161.

Banerjee, S., Critoph, E., and Hart, R. G., "Thorium as a Nuclear Fuel for CANDU Reactors," *Canadian Journal of Chemical Engineering,* vol. 53, p. 291 (1975).

Discusses thorium-cycle neutronics, fuel utilization, and resource requirements.

Bethe, H. A., "The Necessity of Fission Power," *Scientific American,* vol. 234, p. 21 (January 1976).

Argues that a substantial commitment to nuclear power is necessary for economic well-being.

Chang, Y. I. et al., "Alternate Fuel Cycle Options: Performance Characteristics and Impact on Nuclear Power Growth Potential," Argonne National Laboratory report ANL-77-70 (September 1977) (NTIS).

Estimation of fuel utilization characteristics for LWR, SSCR, CANDU, and LMFBR concepts with various fuel cycle options.

The Committee on Nuclear and Alternative Energy Systems of the National Academy of Sciences — National Research Council has, from 1975 to 1977, conducted a study of the need for nuclear power (report expected in 1978).

CONF-740501. "Gas-Cooled Reactors: HTGR's and GCFBR's," topical conference, Gatlinburg, May 7, 1974 (NTIS).

Collection of technical papers on both HTGRs and GCFRs.

Dahlberg, R. C., Turner, R. F., and Goeddel, W. V., "HTGR Fuel and Fuel Cycle Summary Description," General Atomic Company report GA-A12801 (rev. January 1974).

Parameterizes economics of the HTGR versus core composition, burnup, etc.; describes fuel and fuel cycle.

EPRI NP-359. "Assessment of Thorium Fuel Cycles in Pressurized Water Reactors," Electric Power Research Institute report EPRI NP-359 (February 1977) (NTIS).

States effect of thorium cycles on uranium requirements of PWRs.

EPRI NP-365. "Study of the Developmental Status and Operational Features of Heavy Water Reactors," Electric Power Research Institute report EPRI NP-365 (February 1977) (NTIS).

Briefly considers thorium-cycle CANDUs, particularly their relationship to the LWBR program.

ERDA-1. "Report of the Liquid Metal Fast Breeder Reactor Program Review Group," U.S. ERDA report ERDA-1 (January 1975) (NTIS).

An assessment of the need for the LMFBR.

ERDA-1535. "Final Environmental Statement, Liquid Metal Fast Breeder Reactor Program," 3 vols., U.S. ERDA report ERDA-1535 (December 1975), with "Proposed Final Environmental Statement," 7 vols., U.S. AEC report WASH-1535 (December 1974) (NTIS).

A review of the environmental impact of the LMFBR.

ERDA-1541. "Final Environmental Statement, Light-Water Breeder Reactor Program, Commercial Application of LWBR Technology," 5 vols., U.S. ERDA report ERDA-1541 (June 1976) (NTIS).

Among environmental questions, considers uranium utilization of the light-water breeder reactor.

ERDA-76-1. "A National Plan for Energy Research, Development and Demonstration: Creating Energy Choices for the Future, 1976," vol. 1: "The Plan," U.S. ERDA report ERDA-76-1, U.S. Government Printing Office (April 1976).

Includes 1976 projections for the growth of nuclear power until the year 2000.

ERDA-76-107. "Advanced Nuclear Reactors," U.S. ERDA report ERDA–76-107 (May 1976) (NTIS).

A very brief summary of the advanced reactor types that were being considered by ERDA during 1976.

Ford Foundation, "A Time to Choose; America's Energy Future," Report of the Energy Policy Project (Ballinger, Cambridge, Mass., 1974).

Argues that, by increasing the efficiency of energy use, our energy (and electricity) needs will grow much more slowly than they have in recent years.

Ford Foundation/MITRE Corporation, "Nuclear Power: Issues and Choices," Report of the Nuclear Energy Policy Study Group (Ballinger, Cambridge, Mass., 1977).

Projects a low enough nuclear growth and sufficient uranium resources for the LMFBR to be delayed.

Foster, J. S., and Critoph, E., "The Status of the Canadian Nuclear Power Program and Possible Future Strategies," *Annals of Nuclear Energy,* vol. 2, p. 689 (1975).

Treats thorium-fueled CANDUs, as well as those fueled on natural uranium.

GESMO. "Final Generic Environmental Statement on the Use of Recycle Plutonium in Mixed Oxide Fuel in Light-Water Cooled Reactors: Health, Safety, and Environment," 5 vols, U.S. NRC report NUREG-0002 (August 1976) (NTIS).

Examines environmental implications of recycling plutonium in LWRs.

Kasten, P. R. et al., "Assessment of the Thorium Fuel Cycle in Power Reactors," Oak Ridge National Laboratory report ORNL-TM-5565 (January 1977) (NTIS).

Examines the performance of thorium fuels in a variety of reactor types.

Merrill, M. H., "Use of the Low Enriched Uranium Cycle in the HTGR," General Atomic Company report GA-A14340 (March 1977).

Examines the performance of HTGRs with low-enriched fuel.

Perry, A. M., and Weinberg, A. M., "Thermal Breeder Reactors," *Annual Review of Nuclear Science,* vol. 22, p. 317 (1972).

Discusses the potential for thermal reactors to achieve high conversion ratio on the thorium-uranium fuel cycle.

Pigford, T. H., and Ang, K. P., "The Plutonium Fuel Cycles," *Health Physics,* vol. 29, p. 451 (1975).

Examines flows of plutonium in various reactor systems.

Pigford, T. H. et al., "Fuel Cycles for Electric Power Generation," Teknekron report EEED 101 (January 1973, rev. March 1975); "Fuel Cycle for 1000-MW Uranium-Plutonium Fueled Water Reactor," Teknekron report EEED 104 (March 1975); "Fuel Cycle for 1000-MW High-Temperature Gas-Cooled Reactor," Teknekron report EEED 105 (March 1975). These are included in "Comprehensive Standards: The Power Generation Case," U.S. EPA report PB-259-876 (March 1975) (NTIS).

States flow of nuclear materials for a number of reactor types.

Seghal, B. R., Lin, C. L., and Naser, J., "Performance of Various Thorium Fuel Cycles in LMFBRs," Electric Power Research Institute, *EPRI Journal,* vol. 2, p. 40 (September 1977).

Reports nuclear materials requirements for various types of LMFBR fuels.

Till, C. E. et al., "A Survey of Considerations Involved in Introducing CANDU Reactors into the United States," Argonne National Laboratory report ANL-76-132 (January 1977) (NTIS).

Surveys economic (including fuel utilization) and licensing aspects of CANDU reactors.

von Hippel, F., and Williams, R. H., "Energy Waste and Nuclear Power Growth," *Bulletin of the Atomic Scientists,* vol. 32, p. 14 (December 1976).

Argues that ERDA predictions for the growth of nuclear power must involve substantial inefficiency in energy use.

WASH-1097. "The Use of Thorium in Nuclear Power Reactors," U.S. AEC report WASH-1097 (June 1969) (NTIS).

Discusses how thorium-uranium fuel cycles might be used in power reactors.

WASH-1139(74). "Nuclear Power Growth, 1974-2000," U.S. AEC report WASH-1139(74) (1974) (NTIS).

Projects growth of the nuclear power system in the last quarter of this century, as contemplated in 1974.

CHAPTER ELEVEN

The Processing of Nuclear Fuels

COMMERCIAL NUCLEAR POWER in the United States and abroad has recently reached a significant point of transition. Whereas the first and greatest effort in the development of nuclear power was simply to construct a large number of operating nuclear power plants, the next stage must adapt this system to the realities of nuclear resources. These resources are limited and nuclear power must change to reduce its demand for fissile materials from natural deposits.

Growth of commercial nuclear power necessarily entailed development of extraction enterprises and construction of facilities for the production of reactor fuel. Substantial activity is now directed toward the development of processes and facilities for handling irradiated fuel. The purpose of these facilities may be to extract useful uranium, plutonium, and thorium metals from the fuel or to dispose of useless and dangerous radioactive materials, or both. The character of spent fuel facilities can depend strongly on the nature of the nuclear power system being developed.

The principal facility in the "back end" of a complete fuel cycle could be a "fuel reprocessing" plant. As indicated in Figure 11-1, such a plant would dissolve the fuel, separate the resulting solution into one or more fuel and waste fractions, and transform these fractions into desirable form. The fuel materials would eventually be returned to fuel fabrication facilities designed to handle the products of reprocessing. The processed and packaged waste would eventually be disposed of at a carefully chosen site. The facilities for waste processing can, in principle, be considered to be independent of fuel reprocessing activities. However, it is likely that conversion of wastes to a chemical form suitable for disposal would take place on the site of the dissolution and separation facility.

It is possible, of course, to refrain from reprocessing fuels for lengthy periods. This has long been assumed for the Canadian nuclear power system, designed so

162

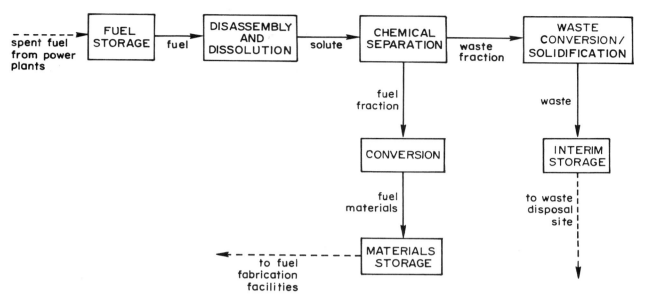

Figure 11-1. SCHEMATIC FUEL PROCESSING FACILITY.
A complete fuel handling operation would include a separations facility and a conversion facility, as well as space for storage of spent fuel and of separated fuel and waste materials.

that the fissile content of spent fuel is very low. Moreover, as it has turned out, light-water reactor fuel in the United States is not being processed. For more advanced reactor systems, it is usually very important that reprocessing occur in order to return bred fissile material to operating reactors. For such systems, frequent reprocessing may be necessary to achieve high conversion or breeding ratios.

As nuclear power develops, other additions or alterations in the nuclear fuel cycle may occur. An obvious one would be the opening of sites for final disposal of nuclear wastes. A possibility that has subtle, but important, implications would be the improvement of enrichment techniques to minimize expense and energy use or to maximize ^{235}U utilization. Finally, alteration of fuel fabrication techniques would be required if recycle of nuclear materials were to occur.

FUEL CYCLES FOR WATER-COOLED REACTORS

The bulk of today's nuclear power system can be operated on a "once through" fuel cycle. Although it is possible, and perhaps even preferable, for nuclear materials to be recycled for the current generation of nuclear power plants, recycle extends resources only a moderate amount. With or without recycle, the annual refueling requirements of present-day plants are large enough to constitute a substantial depletion of resources (see Table 10-2).

There are two well-established commercial reactor types: light-water reactors and heavy-water reactors. The first type has ordinarily been purchased on the presumption that fuel would be reprocessed and that recovered uranium and plu-

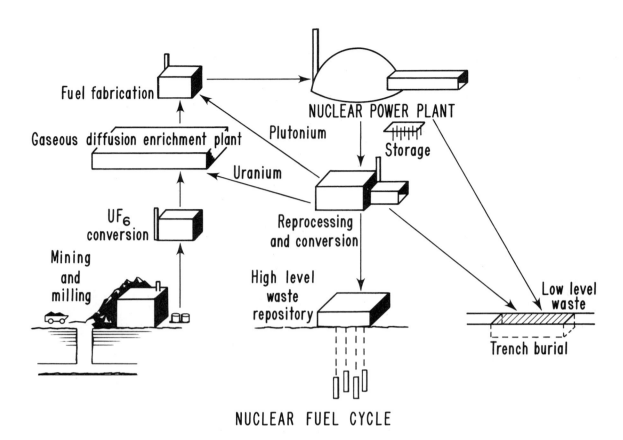

Figure 11-2. MAJOR FACILITIES OF THE LIGHT-WATER REACTOR FUEL CYCLE.
The light-water reactor fuel cycle now includes mining and milling operations, facilities for converting yellowcake to UF$_6$, gaseous diffusion enrichment plants, fuel fabrication plants, and the power plants themselves. A complete fuel cycle which includes recycle of fissile materials would include a reprocessing and conversion plant and a repository for high-level wastes; neither of these facilities now exists. (Figure courtesy of Lawrence Berkeley Laboratory.)

tonium would be recycled to fuel production facilities, as indicated in Figure 11-2. In contrast to LWRs, CANDUs have been operated on the presumption that fuel would be disposed of as waste or stored for use some time in the future. Reprocessing of CANDU spent fuel to recover the 0.27% fissile plutonium content results in as substantial a reduction in lifetime uranium requirements (roughly 4160 tons of U$_3$O$_8$ to 2150) as does recycle of uranium and plutonium for LWRs (roughly 6420 to 4100 tons of U$_3$O$_8$). However, the mass of fuel to be reprocessed is considerably larger in the CANDU.

Stowing away or throwing away spent fuel is also a possibility for light-water reactors. Although the mass of fuel to be processed is only about one-fifth that of a CANDU, economic and other considerations may preclude or postpone reprocessing or recycle. This has been the case in the United States, and a formal decision not to reprocess would lead to the need for long-term spent fuel storage facilities or even waste disposal sites that accept intact fuel assemblies, as discussed later in this chapter.

However, suppliers of light-water reactors have, until recently, proceeded on the assumption that LWR fuel would be reprocessed. Several plants have been designed and built to this end, basically along the lines indicated in Figure 11-3. The only plant that has had any substantial experience in reprocessing fuels is a relatively small plant operated by Nuclear Fuel Services at West Valley, New York. This plant was closed several years ago to improve its equipment and to increase its capacity. However, the operators subsequently decided to close the plant permanently (leaving open the question of who would be responsible for treatment and

Figure 11-3. FLOW DIAGRAM FOR A URANIUM–PLUTONIUM REPROCESSING FACILITY.

In a typical reprocessing operation, the fuel is chopped and dissolved, then uranium and plutonium are extracted from the resulting nitric acid solution. The uranium and plutonium are converted to materials which may be used in the production of fresh fuel. The highly radioactive nitric acid solution is treated to solidify the wastes, which are then available for disposal. (Figure reproduced from ERDA-1541.)

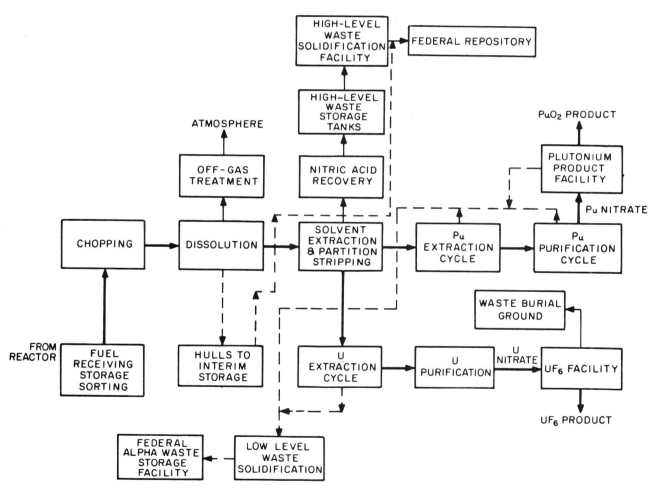

disposition of the radioactive wastes from its earlier period of operation). A second commercial reprocessing plant was largely completed at Morris, Illinois, by General Electric. Plans for opening this plant were abandoned when it was discovered that, with its particular design features, it would not operate properly. Finally, at Barnwell, South Carolina, Allied General Nuclear Services has built a plant, designed along more conventional lines,[1] which would be ready for operation but for the fact that it has no facilities for conversion of the fuel or waste fractions to solid form (see Figure 11-1). Until detailed criteria for these conversions are developed, it is unlikely that this plant can begin operation. The Barnwell facility is designed to reprocess 1500 metric tons of fuel per year, permitting it to serve about 50 1000-MWe LWRs. The United States nuclear capacity is rapidly exceeding this amount and, presuming reprocessing is begun, additional reprocessing plants would be required. Other companies have taken preliminary steps to secure licenses to build such plants.

However, it is apparent that a number of questions will have to be resolved before reprocessing can occur. From the point of view of utilities, the major question is economic: Does recovery of fissile material in spent fuel justify the expenditure for reprocessing? (In this comparison, the relative costs of fuel storage and waste disposal modes have to be considered.) However, additional questions remain: What are the health and safety implications of fuel reprocessing? How is reprocessing connected with possible diversion of nuclear materials to weapons purposes? The latter question is the subject of the next chapter. The first question is largely a matter of how stable a method can be found for final disposal of reprocessing plant wastes (see later section), assuming that the reprocessing plant itself is designed to suitable health and safety standards. (It is interesting to note, though, that the relatively small average world population dose that would occur from a substantial nuclear power system is largely attributed to the release of gaseous radioactivity at the reprocessing plant — unless special measures are taken to retain these gases.) In this connection, it is important to note that use of recycled material in a fuel fabrication plant would require special precautions for the protection of workers (see next section).

Were reprocessing to occur, the fuel and waste fractions, which would probably be separated as nitrate solutions (see Appendix F), would presumably be converted to other forms. The recovered uranium would be converted to uranium hexafluoride for return to an enrichment plant, as indicated in Figure 11-2. If plutonium were to be recycled, it would be converted to plutonium oxide; if long-term storage or disposal were chosen, another form might be selected. Finally, the waste fractions can be handled as noted later in this chapter. Light-water reactor fuel cycle materials flows, assuming reprocessing and uranium recycle but with and without plutonium recycle, are given in Appendix F. An alternative choice is, of course, not to reprocess at all.

Many countries are proceeding vigorously with efforts to begin reprocessing. The uranium and plutonium in spent fuel represents a valuable resource for both

[1] A brief description of typical plant processes is included in Appendix F.

TABLE 11-1.
Amount of Fuel Reprocessed Annually as a Function of
Reactor Type

Reactor	Approximate Reprocessing Needs (metric tons of fuel metal per year for 1000 MWe at 75% capacity factor)
LWR	26
Thorium PWR	26
LMFBR	20
HTGR	8
CANDU	114
CANDU (Pu recycle)	60
Thorium CANDU (high burnup)	30
Break-even CANDU	60 – 90

light-water reactors and fast breeder reactors. Most countries that depend heavily on commercial nuclear power are strongly inclined to exploit this resource. Moreover, many countries regard reprocessing to be necessary to cast nuclear wastes into a form suitable for final disposal. The United States is moving more cautiously in this matter. A genuine consideration is not only how reprocessing might effectively extend uranium resources and prepare wastes for disposal, but how it would be implemented for more advanced nuclear systems, particularly with a view to minimizing the tendency to weapons proliferation.

ADVANCED REACTORS AND FUEL RECYCLE

Advanced reactors, in the usual sense, are designed to improve the conversion ratio over that of current reactor types, preferably to the point where at least as much fissile material is produced as is consumed. The conversion ratio is improved by careful attention to minimizing neutron absorptions by materials other than uranium, plutonium, or thorium. Frequent reprocessing serves to prevent the buildup of poisons that would rob the system of neutrons. Moreover, it is inevitable that irradiated fuel removed from a reactor will have a substantial fissile content, particularly in breeder reactors, and reprocessing is necessary to recover this material. For these reasons, reprocessing is typically presumed when discussing breeders and near-breeders. Because of the special interest in minimizing demands on uranium resources, the choice of whether to reprocess is not nearly as free as for current water-cooled reactors.[2] Figure 11-1 applies equally well to reprocessing of fuel from advanced reactors and from LWRs. However, the detailed design of the reprocessing plant and of other fuel cycle facilities may differ on several counts.

The basic reprocessing steps may change slightly from one fuel cycle to another. The precise fuel makeup, and even the fissile content, will differ. If the

[2] Chapter 12 discusses the potential of "once-through" fuel cycles for reducing the probability of proliferation.

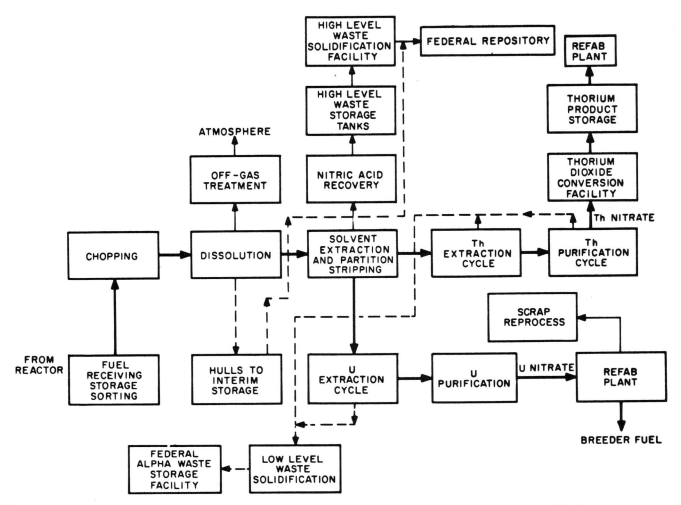

Figure 11-4. FLOW DIAGRAM FOR A THORIUM–URANIUM REPROCESSING FACILITY. The basic operations in reprocessing fuel from the thorium-uranium fuel cycle are similar to those used for the uranium plutonium cycle. A significant difference arises merely from the fact that the principal fissile isotope from thorium-uranium fuel is ^{233}U, not ^{239}Pu, so that the process would be very careful not to lose uranium. (Figure reproduced from ERDA-1541.)

fissile content is very high at any point in the fuel stream, precautions will have to be taken to prevent assembly of a critical mass at any point in the fuel stream. A spontaneous fission chain reaction would seriously endanger those nearby, even though an explosive reaction could not occur accidentally (see next chapter).

As for the chemical separation and conversion techniques, those for LWR fuel (Figure 11-3) are equally applicable to ordinary LMFBR fuel which, after all, is also uranium and plutonium oxide.[3] Near-breeder reactors, on the other hand, inevitably include thorium, as may certain types of fast breeders. Separation of the fuel

[3] However, carbide fuels are also considered.

materials into the three fuel streams, uranium, plutonium, and thorium, would be slightly more difficult than the LWR process. More often, though, the fuels would be separated so that, in addition to a uranium-plutonium process stream (Figure 11-3), there would be one for thorium-uranium (Figure 11-4). Except for this factor, the separation process for fuels from most important reactor types is similar because the fuel is typically a dioxide in a Zircaloy or stainless steel can. The obvious exceptions are gas-cooled reactors with carbon moderators; HTGR fuel blocks require crushing and burning to free the ceramic fuel pellets and these pellets may be carbides or oxides.

The radionuclide content of various irradiated fuels differs in significant ways. From the point of view of difficulty of reprocessing, the cooldown time after removal from the reactor is extremely important. Minimizing the fissile inventory in the fuel cycle is a strong incentive for reprocessing after only a short time, but this means processing highly radioactive fuel. The total radioactive content of the fuel does not vary greatly from one reactor to another (even though individual radio-nuclide concentrations may vary substantially, depending on power density, burn-up, and other factors). A longer wait gives certain short-lived isotopes a chance to decay (see Table 3-1). Some of the volatile fission products, such as ^{131}I, have biological significance, and early processing can increase occupational and general public exposures. The nonvolatile radioactivite species also need attention, particu-larly because of the heat associated with their decay.

It is ordinarily assumed that LWR fuel would cool for at least six months prior to processing. For an LMFBR, however, it would be advantageous to decrease the cooling period as much as possible, even to 30 days, to reduce inventory and thus the doubling time (see Chapter 10). This may not, in fact, be so critical in a slowly growing economy. But even with the near-breeder systems, which are suit-able for such an economy, minimizing the fuel cycle inventory reduces the demand on uranium resources.

The impact of recycling fuel materials goes beyond the reprocessing plant in a manner that depends on the fuel cycle employed. The uranium-plutonium cycle yields recycled uranium with an isotopic composition (see Table 11-2) that does not require radiological precautions more severe than those for handling natural uranium. However, the recycle of plutonium requires substantial precautions, largely in the nature of "glove boxes," in which the material is handled by person-nel who insert their arms into long gloves sealed into the walls of transparent boxes. These boxes are designed to confine the radioactive material. The radiation from plutonium is so short ranged that no special shielding is required.

In contrast to this, reliance to any extent on thorium and ^{233}U imposes an additional burden in connection with small amounts of ^{228}Th and ^{232}U in the thorium and uranium fractions extracted at the reprocessing plant. These isotopes have half-lives of intermediate length (2 years and 70 years, respectively), and they both lead to daughters that have a moderate probability of emitting unusually penetrating radiation. As a result, fuel fabrication facilities would have to be de-signed with thick shielding and remote handling in order to protect workers at the facility. This would ordinarily be the case at the reprocessing plant, anyway, but is

TABLE 11-2.
Typical Isotopic Composition of Extracted Uranium and Plutonium[a]

LWR with Uranium Recycle

	Uranium[b]				Plutonium[b]		
	Percent	kg	Ci		Percent	kg	Ci
234	0.01%	3	19.2	236	1×10^{-4}%	2×10^{-4}	1.3×10^2
235	0.83	213	0.5	238	2.4	6	1.0×10^5
236	0.44	113	7.1	239	58.4	142	8.7×10^3
238	98.7	25,422	8.4	240	24.0	59	1.3×10^4
	100%	25,751	35.2	241	11.2	27	2.8×10^6
				242	3.9	10	3.7×10^1
					100%	244	1.2×10^5 (α)
							2.8×10^6 (β)

LWR with Natural Uranium plus Recycled Plutonium

	Depleted Uranium[b]				Extracted Plutonium[b]		
	Percent	kg	Ci		Percent	kg	Ci
234	0.01%	1.7	10.2	236	0.0%	0.0	2.2×10^2
235	0.33	81.8	0.2	238	4.2	41	6.9×10^5
236	0.08	19.8	1.2	239	38.7	380	2.3×10^4
237	0.00	0.0	388.0	240	27.5	270	6.0×10^4
238	99.59	24,900	8.3	241	18.1	178	1.8×10^7
	100%	24,992	408	242	11.5	113	4.4×10^2
					100%	982	7.8×10^5 (α)
							1.8×10^7 (β)

LMFBR with Depleted Uranium plus Recycled Plutonium and Uranium

	Extracted Plutonium[b]		
	Percent	kg	Ci
236	2×10^{-6}%	3×10^{-5}	1.7×10
238	0.1	1.4	2.4×10^4
239	71.7	1,460.6	9.0×10^4
240	25.1	510.8	1.1×10^5
241	2.4	48.4	4.9×10^6
242	0.8	15.5	6.0×10
	100%	2,037	2.3×10^5 (α)
			4.9×10^6 (β)

HTGR with Recycled 233U and Once-Recycled 235U Feed

	Extracted Uranium Feed[b] (once through, from 93% 235U feed)				Bred Uranium[b] (recycled from thorium)		
	Percent	kg	Ci		Percent	kg	Ci
234	0.1%	0.1	0.3	232	0.0%	0.1	2,307
235	21.8	17.4	0.0	233	55.4	187.6	1,778
236	55.8	44.6	2.8	234	23.2	78.7	487
237	0.0	0.0	0.9	235	9.5	32.3	0.1
238	22.3	17.8	0.0	236	11.5	39.0	2.5
	100%	79.8	4.2	237	0.0	0.0	0.6
				238	0.3	1.0	0.0
					100%	338.8	4,575

a. Yearly masses and radioactivity from 1000-MWe plant (80% plant factor), assuming 150-day decay time, except for LMFBR, which assumes 30 days; from Pigford and Ang.
b. For each isotope, percent by mass, mass in kg, and radioactivity in Ci are given.

in contrast to the fabrication plants associated with the uranium-plutonium fuel cycle, where the oxides can be handled at close range without subjecting workers to large radiation doses. In a thorium-uranium system, it is almost certain that great value will be attached to quick recycle of ^{233}U, so that storage to permit decay of the ^{232}U would not be practical and recycle facilities would have to be designed to cope with this penetrating radiation. On the other hand, it would not be difficult, considering the availability of thorium and the two-year half-life of ^{228}Th, to store recycled thorium long enough to permit a substantial decrease of the radiation emitted.

In other respects, the recycle facilities for the thorium-uranium fuel cycle do not differ greatly from those of the uranium-plutonium cycle. The fissile isotope of interest, ^{233}U, is somewhat less toxic (as judged by prevalent radiation protection standards) than recycled plutonium, but any alteration of handling techniques that would be associated with this difference is overwhelmed by the changes required by the presence of ^{232}U. As for the chemical processes used for separation and conversion, slight alterations will be required because of the presence of thorium and because, in this case, it is the uranium that is the valued species.

For some of the thorium-uranium fuel cycles, the cost of reprocessing and fabrication can be substantially more than the comparable costs in the uranium-plutonium fuel cycle. This difference arises simply because, per unit of energy generated, the actual bulk of material reprocessed may be greater in thorium-based near-breeder cycles. A clear indication of this is given by the low burnup permitted by these fuel cycles if break-even is to be achieved, i.e., about 10,000 MWd/Te. All told, the fuel cycle costs for near-breeders can become a substantial part of the total cost of generating electricity. However, the power plant capital cost remains the largest component in the generation cost. According to Atomic Energy of Canada Limited, the total generation cost from a break-even CANDU fuel cycle is only about 20% greater than current CANDU costs. However, the cost increase depends greatly on fuel reprocessing costs and recycle fabrication costs, and these are uncertain at this time.

At this point, it is appropriate to note that consideration and development of advanced reactor types and their associated fuel cycles have generally focused primarily on resource utilization and economic questions. In addition, substantial attention has been given to health and safety aspects of these systems, including safety design, radiation protection needs, etc. However, a broad area that has received, until recently, rather cursory attention is the difficulty of preventing commercial nuclear materials from being used for weapons. Substantial attention to prevention of diversion and proliferation could favor particular reactor/fuel types and could lead to substantial alteration of reactor and fuel cycle design.

WASTE PROCESSING AND DISPOSAL

The operation of a nuclear power plant generates large amounts of radioactivity, primarily from reactions in the fuel material. Fissions yield various medium-weight elements, particularly halides, noble gases, and rare earths. Neutron capture typically leads to the "actinides," a chemical class that includes uranium,

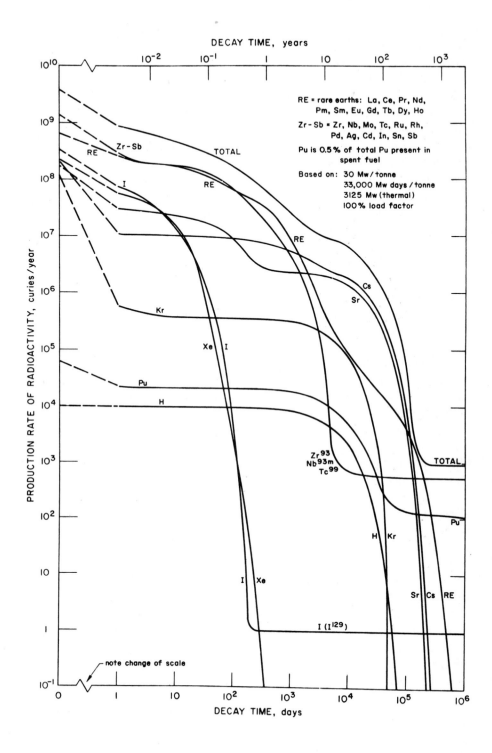

plutonium, and other transuranics. In addition, structural materials, including fuel cladding, can be activated by interaction with neutrons and gamma rays; such radiation can also induce reactions in the coolant or moderator and in other structural materials, such as the reactor vessel walls. It is, however, the fission products and actinides that comprise the bulk of the power plant's radioactivity. Table 3-1 indicated a representative radioactive inventory for an operating reactor. Since most of the power plant inventory is directly associated with the core itself, this radioactivity is removed with the spent fuel. Short-lived isotopes rapidly decay to very low levels, but substantial amounts of radioactivity remain for a long time, so that the irradiated fuel assemblies constitute a substantial waste handling and disposal problem. (See Figure 11-5.)

For any fuel cycle in which spent fuel is reprocessed, these radioactive species are removed from the fuel assemblies to either a fuel or waste stream (Figure 11-1). An alternative, of course, is not to reprocess, but to store or dispose of fuel directly. This possibility is discussed at the end of this section. Presuming reprocessing, uranium, plutonium, and thorium are extracted from an acidic solution produced in the course of dissolving the fuel. The remaining solution is a high-level liquid waste (HLLW) which must be isolated from the environment because of its extremely high radioactive content. Other waste streams also result from reprocessing, the most notable of which are the cladding hulls which remain after the fuel is dissolved. For LWRs, on the order of 15 metric tons of hulls are generated per gigawatt-year of operation. These and other radioactive wastes, from both the reprocessing plant and elsewhere, require careful disposition. However, because of its greater radioactive content, the high-level waste has received most attention and is the principal subject of this section.

The high-level waste has two principal classes of radioactive constituents: fission products and actinides. Nitrates of these species are immersed in a large, acidic liquid volume, on the order of 1000 liters per metric ton of fuel dissolved. Depending on the details of dissolution and extraction, the volume may be somewhat more or less than this,[4] but usually it can be reduced to about 400 liters/Mg, prior to further treatment. Thus about 10,000 to 20,000 liters of concentrated high-level wastes are generated per gigawatt-year.

Not surprisingly, it is Nuclear Regulatory Commission policy that, at any reprocessing plant, the high-level wastes would have to be reduced to solid form. However, detailed criteria have not been developed with regard to either the nu-

---

[4] Many of the wastes from U.S. military programs are more dilute than the commercial liquid wastes would be. Because much of the military wastes have been stored in this liquid form, many tank storage units have been constructed on military reservations.

Figure 11-5. DECAY OF RADIOACTIVE WASTES FROM LIGHT-WATER REACTORS. The radioactivity associated with various waste species from light-water reactors is shown as a function of time after removal from reactors. Many species decay with short enough half-lives for the total radioactivity to decrease by a factor of a million within a few hundred years. The principal radionuclides remaining after that time are isotopes of plutonium, zirconium, niobium, and technetium. (Figure reproduced from Pigford, "Fuel Cycles. . . .")

clides to be solidified or the form they would take. These specifications may depend on the method of long-term or final disposal chosen.

It may be advantageous to "partition" the radioactivity into separate fractions containing fission products and actinides. The advantage could arise from the fact that, for the most part, the fission products have short half-lives and the actinides have longer ones. Note the behavior of various species, particularly plutonium, that is indicated in Figure 11-5 for uranium-fueled LWRs. It is therefore the actinides, as a class, which require a disposal method that is particularly stable. It would also be useful to remove some of the longest lived fission products for similar disposal.

The amount of actinides in high-level wastes depends somewhat on the form of the fuel cycle. Recycle of plutonium results in production of more of the higher actinides, particularly americium and curium. However, it also reduces the amount of plutonium requiring storage or disposal. As a result, direct disposal of irradiated LWR fuel relegates a larger amount of actinides, particularly as measured in terms of toxicity, to storage than does an LWR with plutonium recycle (assuming only about 1% of the plutonium is lost to wastes). Other types of fuel cycles would have somewhat different amounts of actinides, but it is fair to say that the amounts to be disposed of, and the hazard they represent, do not depend strongly on the fuel cycle chosen. As a result, the requirements for treating liquid wastes from reprocessing can be considered relatively independent of particular fuel and recycle schemes.

The liquid wastes may be solidified by a variety of methods, many of which involve "calcining" the wastes to obtain a dry powder as granules. Calcination, which is basically just heating liquid droplets in air to a dry form, may be accomplished in a number of ways. One of the more favored solidification methods immediately incorporates the calcine solids into glass. This is accomplished as indicated in Figure 11-6, which shows a combined spray calcination and in-canister melting process. Material for making glass (frit) is added to the calcine as it falls into a heated canister. There, glass is formed. The canister is then disposed of. Glass is a particularly inert substance, which is why it is chosen for immobilizing the high-level wastes. Spray calcination with in-can melting is among the more highly developed solidification techniques; a pilot solidification unit almost as large as required for a Barnwell-sized reprocessing plant has been operated. Figure 11-7 illustrates a technique that is also regarded favorably: calcine or liquid wastes are fed into a melter where the materials are heated with frit to form a glass melt that overflows into a waste canister, the unit for final disposal.

The waste disposal technology that has not been tested on a pilot scale is the actual long-term disposal of solidified wastes. Various alternatives have been considered, including sea-bed disposal, and shooting the actinides into space. However, the method that has been most highly developed is disposal in geologic formations, such as salt beds and crystalline rock (granite). In any such disposal, it is important that no groundwater be available for leaching or carrying away radioactivity that may escape from canisters. Salt formations fall into this category (although care must be taken that unrelated human activities do not alter the stability of the formation). Granitic disposal is arousing increasing interest. In either case, much

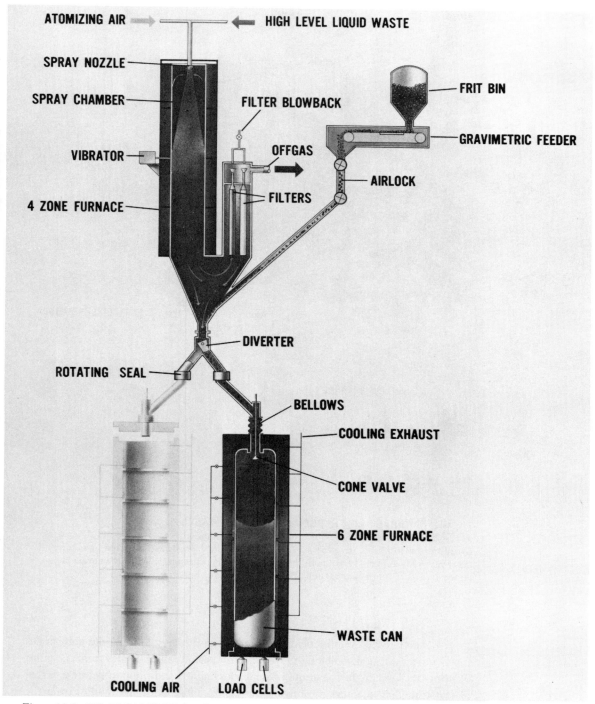

Figure 11-6. SPRAY CALCINER/IN-CANISTER MELTER FOR WASTE SOLIDIFICATION.
In the upper portion of this waste solidification system, liquid high-level wastes are blown into
a furnace region, where droplets are heated to a dry form. This material then falls, along with
glass-making frit, into a metal canister. This canister is surrounded by a melter furnace, which
forms glass of the material in the canister. This canister is then deposited at a waste disposal
site. (Figure courtesy of Battelle Pacific Northwest Laboratory.)

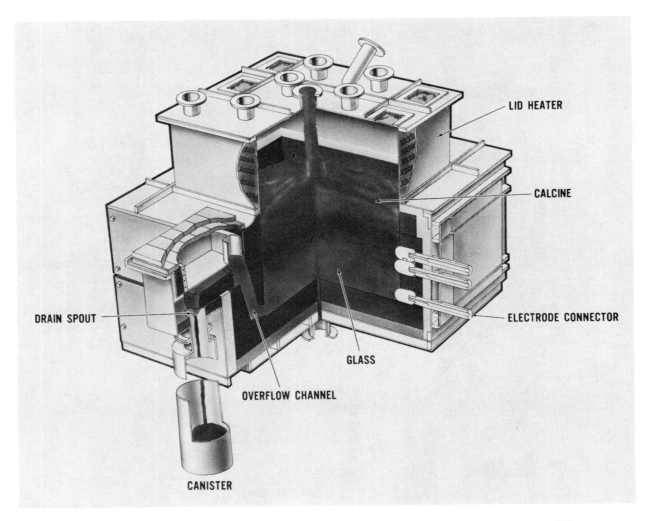

Figure 11-7. DIRECT LIQUID FEED MELTER FOR WASTE SOLIDIFICATION.
This solidification system is designed to accept either liquid high-level wastes or calcine produced from such wastes. In either case, the waste fed into the system is heated together with other materials to form a glass melt; this overflows into a canister, which is then available for disposal. (Figure courtesy of Battelle Pacific Northwest Laboratory.)

work remains to be done in identifying candidate sites, determining suitable modes for disposal, and understanding movement of released radionuclides from the point of disposal. A major factor in the waste disposal question is how to ensure a degree of isolation that adequately protects the environment (and the public) from the escape of stored radioactivity. Figure 11-8 indicates the location of various types of geologic formations in the United States.

For some time, federal agencies have been developing plans for the form of a final disposal site. An artist's conception of one possibility is shown in Figure 11-9. The high-level waste canisters are dropped down a shaft, then moved to burial rooms, where they are deposited in holes drilled in the geologic formation. Another shaft is available for workers who perform these operations (as indicated, this shaft may also be used for transmission of low-level wastes to a burial room). If development of a final disposal site is deferred, wastes may be stored on the surface as an interim measure. Figure 11-10 shows a possible mode of surface storage.

In contrast to the discussion in this chapter, neither fuel reprocessing nor waste solidification need occur. Presuming the nuclear power system is designed to forego use of the remaining fissile material, spent fuel may be stored or deposited directly. Most of the reaction products are, after all, immobilized in the ceramic fuel pellets (at least in oxide-fueled reactors), and they need not be disturbed. In principle, the fuel assembly itself can be encapsulated and buried. This would, in

Figure 11-8. POSSIBLE GEOLOGIC FORMATIONS FOR WASTE DISPOSAL IN THE UNITED STATES.
Salt deposits have been the geologic formations given the most attention for waste disposal in the United States. Argillaceous formations include shales in various parts of the country. The crystalline formations include granite, material that is also being seriously investigated as a suitable matrix for waste disposal. (Figure reproduced from ERDA-76-162.)

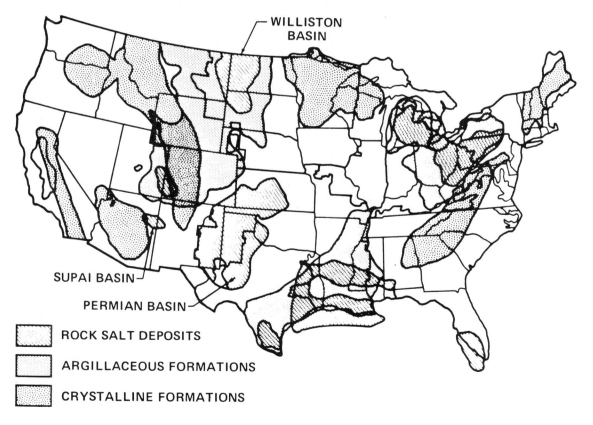

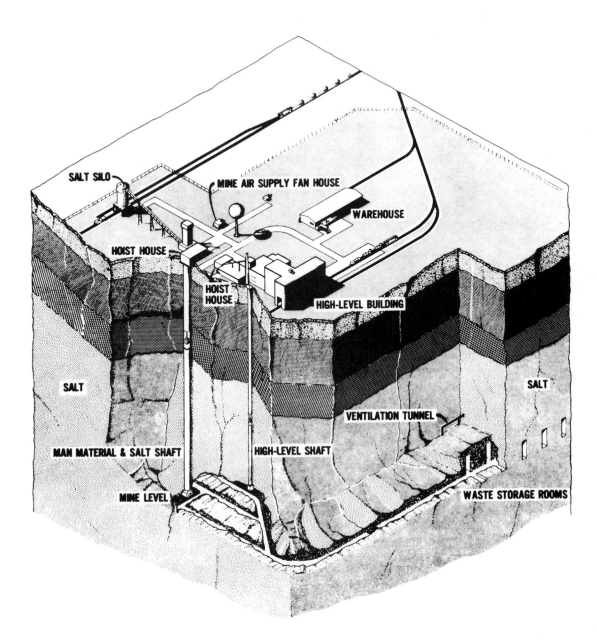

Figure 11-9. POSSIBLE FORM FOR FINAL WASTE DISPOSAL SITE.
The figure shows an artist's conception of a waste disposal site in which the wastes are deposited in a salt formation. Numerous service buildings are on the surface, with shafts leading down to burial rooms. (Figure reproduced from NUREG-0116.)

fact, avoid many of the complications of waste solidification. Alternatively, the fuel can simply be stored for a long time at a so-called "retrievable surface storage facility." Fuel assemblies can be placed in casks surrounded by concrete, and cooled with either forced or natural convection. This form of storage would be similar to that indicated in Figure 11-10. An advantage to retrievability is that, if necessary, the fuel can later be reprocessed, either to recover fissile material or to render wastes into a more suitable form.

ENRICHMENT TECHNOLOGIES

The enrichment plant is a fuel cycle facility that we have heretofore neglected. Enrichment is associated largely with light-water reactors and also with initial feed requirements of some advanced thermal reactors. However, recent developments in enrichment technologies have raised, on the one hand, the possibility that uranium resources may be more effectively utilized and, on the other, the spectre of increased accessibility of nuclear weapons materials.

For any process intended to separate two isotopes or, putting it another way, to alter the ratio of two isotopes in a mixture, the effectiveness of the process may

Figure 11-10. SEALED STORAGE CASK.
Waste canisters may be stored in a steel cask and surrounded by concrete. The unit may be designed to be natural convection cooled, with low radiation levels at the exterior. Such a unit may sit on the surface, rather than be buried. A variation of it may be used for spent fuel storage. (Figure reproduced from NUREG-0116.)

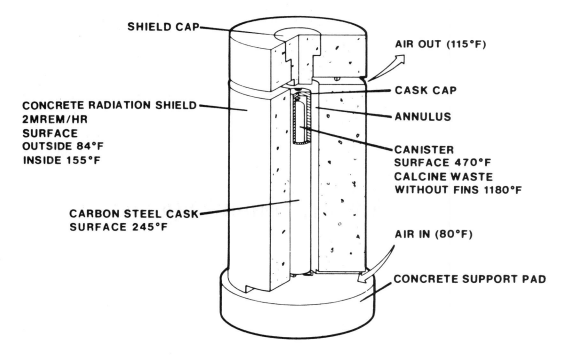

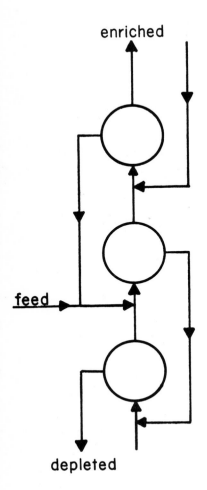

enriched

feed

depleted

Figure 11-11. SCHEMATIC
ENRICHMENT CASCADE.
Each of the circles in the figure represents a single stage in a uranium
enrichment cascade. Material flows
into each stage, and two streams are
produced, one enriched in ^{235}U, the
other depleted. The figure shows the
three central stages of an enrichment
cascade, including the one into which
the feed material is introduced.

be expressed in terms of a separation factor, f. Designating the ratio of the two
isotopes as r, the process separates a quantity of material into an enriched portion
and a depleted portion, wherein the new ratios are r_e and r_d, respectively. The
separation factor is defined to be the ratio $f = r_e/r_d$. In the usual case of ^{235}U and
^{238}U, then, the ratio starts as r = 0.007 (i.e., 0.7% ^{235}U), and the question is, How
effectively can a given technique raise this ratio, "enriching" the uranium in ^{235}U?
If each process stage has only a small separation factor, many stages will be required
to achieve the desired enrichment. The stages would, in fact, be cascaded so that, at
each stage, the enriched product feeds a higher stage, and the depleted product
feeds a lower one. As seen in Figure 11-11, the feed of any given stage is then the
depleted fraction from the next higher stage plus the enriched fraction of the next
lower stage. The cascade is designed so that natural uranium feed is supplied at a
central stage; the enriched product is drawn from one end of the cascade while
"tails" are drawn from the other. The difference in the enriched and depleted
isotopic ratios depends only on the number of stages in the cascade.

As it turns out, for the gaseous diffusion technique, in which UF$_6$ is passed
through a porous barrier (see Figure 11-12), the separation factor can be shown to
equal the square root of the molecular mass ratio, which for ^{235}UF$_6$ and ^{238}UF$_6$
is $\sqrt{352/349} = 1.0043$. Evidently, many stages are required to substantially change
the percentage of ^{235}U, particularly since the actual separation factor is smaller
than the theoretical value of 1.0043. As a result, about a thousand stages (within a
factor of 2) are necessary to raise the ^{235}U content to that necessary for LWRs (2%
to 4%), assuming that the tails percentage is in the vicinity of 0.2% to 0.3% ^{235}U.
Because the stages are sizable, this does not result in a small facility. (To raise the
enrichment further to moderate or high levels, say 50% to 95%, requires at least a
thousand more stages.) Substantial investments are required to build such large
facilities, and substantial energy is required (see end of Chapter 2) for heating the
UF$_6$ to a gaseous state and pumping it through the porous barriers.

The amount of effort required to complete a certain enrichment task is
typically measured in terms of "separate work units," with units of kilograms. This
is literally a relative measure of the material pumped through various stages. A
specific reactor type will require a yearly feed of uranium at a specified enrichment,
and, assuming the tails assay is also specified, this will require a definite amount of
separative work. See, for example, Table 10-1. In 1977, separative work exceeded a
cost of $50/kg for new contracts. The cost has been rising rapidly, largely because
of increases in the cost of energy to operate diffusion plants and also because of
changes in U.S. pricing policy.[5]

Other enrichment technologies are becoming available. Plants using high-
speed centrifuges (Figure 11-13) are now being designed and built, particularly in
Europe. This method also uses the mass difference between ^{235}U and ^{238}U to
effect a separation, but the separation factor per stage is much greater (perhaps

---

[5] U.S. gaseous diffusion plants, built originally as part of the weapons program, are
operated by contractors to the Department of Energy.

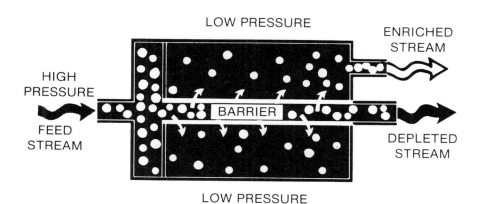

Figure 11-12. SCHEMATIC OF A GASEOUS DIFFUSION STAGE.
The gaseous diffusion technique enriches uranium by permitting hot uranium hexafluoride gas to diffuse through a porous barrier. The lighter molecules, those bearing ^{235}U, have a slightly higher probability of diffusing through the barrier than those bearing ^{238}U. The material that passes through the barrier is drawn off as an enriched stream; the remnant is the depleted stream. (Figure courtesy of U.S. DOE.)

1.25 to 2) than in gaseous diffusion. As a result, fewer stages are required to produce reactor fuel, on the one hand, or weapons material, on the other. Centrifuge enrichment promises to cost less and use considerably less energy than gaseous diffusion. Other mass-based separation techniques are available or under development. One, for example, pumps UF_6-bearing hydrogen around a pair of blades, one of which splits the gas stream into components which are enriched and depleted in ^{235}U. The only other technique that appears to have reached the stage of commercialization is chemical enrichment, in which slight differences in the chemical reactivity are exploited to effect a separation. France is planning to build such an enrichment facility.

However, the technique that has caused the most excitement and concern — laser enrichment — depends only indirectly on mass differences. In this technique, powerful lasers produce intense monochromatic light beams to ionize either atoms of the selected uranium isotope or molecules bearing the selected isotope. Once this occurs, the two isotopes are relatively easily separated by chemical or physical means. Substantial efforts are being expended in the United States and abroad to develop a successful laser enrichment technique. The possible advantages are a large separation factor, lower cost, lower energy requirements. The large separation factor may make one-pass enrichment possible and may minimize the amount of ^{235}U that is largely wasted by being left in the tails. Even ^{235}U present in existing stockpiles of depleted uranium could become a substantial resource.

It is clear, then, that laser enrichment could have important implications for both light-water reactors, requiring low enrichment, and for advanced reactors, some of which require highly enriched material. And is it even conceivable that, with large separation factors, recycled fissile materials may undergo isotope separation to remove less desirable isotopes? However, regardless of the potential advantages to commercial nuclear power, successful and efficient laser enrichment could exacerbate the problem of weapons proliferation by making available a technique that can succeed, on a relatively small scale, in producing weapons grade nuclear materials.

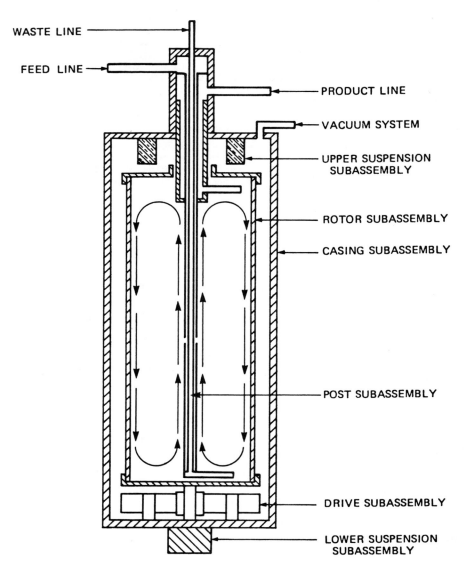

Figure 11-13. SCHEMATIC OF A GAS CENTRIFUGE STAGE.
Pressure differences in a centrifuge can effect isotopic separation. In the device shown, the rotor assembly is spun at high speed to bring about this separation. Uranium hexafluoride flows within the assembly as indicated. The centrifugal action causes selective diffusion of ^{238}U from the upward flow to the outside of the rotor interior and of ^{235}U from the downward flow toward the inside. As a result, the upper regions are enriched in ^{235}U. (Figure reproduced from ERDA-1543.)

Bibliography — Chapter Eleven

"NTIS" indicates report is available from: National Technical Information Service, U.S. Department of Commerce, 5285 Port Royal Road, Springfield, VA 22161.

APS-1977. L. C. Hebel et al., "Report to the American Physical Society by the Study Group on Nuclear Fuel Cycles and Waste Management" (July 1977). To be published as supplement to *Reviews of Modern Physics,* vol. 50.

Discusses alternatives for reprocessing, recycle, and waste management, including their environmental implications.

Bebbington, W. P., "The Reprocessing of Nuclear Fuels," *Scientific American,* vol. 235, p. 30 (December 1976).

Describes the processes involved in separating irradiated nuclear fuel into fuel and waste fractions.

Casper, B. M., "Laser Enrichment: A New Path to Proliferation," *Bulletin of the Atomic Scientists,* vol. 33, p. 28 (January 1977).

Argues that laser enrichment may so exacerbate the proliferation problem that development should be suspended for the time being.

Cohen, B. L., "The Disposal of Radioactive Wastes from Fission Reactors," *Scientific American,* vol. 236, p. 21 (June 1977).

Describes how radioactive wastes might be stored.

————, "Hazards from Plutonium Toxicity," *Health Physics,* vol. 32, p. 359 (1977).

Argues that plutonium, as used in nuclear power, does not constitute a substantial hazard.

CONF-76-0701. "Proceedings of the International Symposium on the Management of Wastes from the LWR Fuel Cycle," Denver, July 11-16, 1976 (NTIS).

Collection of technical papers on management of LWR wastes.

Cowan, G. A., "A Natural Fission Reactor," *Scientific American,* vol. 235, p. 36 (July 1976).

Describes discovery of the site of a natural fission reactor, active 2 billion years ago, in the Republic of Gabon.

Dahlberg, R. C., Turner, R. F., and Goeddel, W. V., "HTGR Fuel and Fuel Cycle Summary Description," General Atomic Company report GA-A12801 (rev. January 1974).

Describes fuel cycle operations associated with HTGRs.

ERDA-1535. "Final Environmental Statement, Liquid Metal Fast Breeder Reactor Program," 3 vols., U.S. ERDA report ERDA-1535 (December 1975), with "Proposed Final Environmental Statement," 7 vols., U.S. AEC report WASH-1535 (December 1974) (NTIS).

Environmental statement on LMFBR program, including the implications of fuel reprocessing.

ERDA-1541. "Final Environmental Statement, Light Water Breeder Reactor Program, Commercial Application of LWBR Technology," 5 vols., U.S. ERDA report ERDA-1541 (June 1976) (NTIS).

Discusses reprocessing requirements for uranium-plutonium and thorium-uranium fuels.

ERDA-1543. "Final Environmental Statement, Expansion of U.S. Uranium Enrichment Capacity," U.S. ERDA report ERDA-1543 (1976) (NTIS).

Considers need and impact of increasing U.S. enrichment capacity, using either gaseous diffusion or centrifuge units.

ERDA-76-43. "Alternatives for Managing Wastes from Reactors and Post-Fission Operations in the LWR Fuel Cycle," 5 vols., U.S. ERDA report ERDA-76-43 (May 1976) (NTIS).
  Describes the various options available for radioactive waste management.

ERDA-76-162. "The Management and Storage of Commercial Power Reactor Wastes," U.S. ERDA report ERDA-76-162 (1976) (NTIS).
  A brief summary document, based on ERDA-76-43.

ERDA-77-75. "Light Water Reactor Fuel Reprocessing and Recycling," U.S. ERDA report ERDA-77-75 (July 1977) (NTIS).
  Considers impact of reprocessing and recycle for light-water reactors.

"Final Environmental Statement, 40 CFR 190: Environmental Radiation Protection Requirements for Normal Operations of Activities in the Uranium Fuel Cycle," 2 vols., U.S. EPA report EPA 520/4-76-016 (November 1976).
  States background for emission standards for the uranium-fueled LWR fuel cycle.

Gall, N., "Atoms for Brazil, Dangers for All," *Bulletin of the Atomic Scientists,* vol. 32, p. 4 (June 1976).
  Discusses significance of Brazilian acquisition of fuel cycle facilities.

GESMO. "Final Generic Environmental Statement on the Use of Recycle Plutonium in Mixed Oxide Fuel in Light Water Cooled Reactors: Health, Safety, and Environment," 5 vols., U.S. NRC report NUREG-0002 (August 1976) (NTIS).
  Environmental statement on reprocessing and recycle of plutonium in light-water reactors.

Glackin, J. J., "The Dangerous Drift in Uranium Enrichment," *Bulletin of the Atomic Scientists,* vol. 32, p. 22 (February 1976).
  Describes enrichment process to be used in Brazilian facilities.

Gofman, J. W., and Tamplin, A. R., *Poisoned Power* (New American Library, New York, 1974).
  Strong criticism of the environmental impacts of nuclear power.

Krass, A. S., "Laser Enrichment of Uranium: The Proliferation Connection," *Science,* vol. 196, p. 721 (1977).
  Examines the question whether laser enrichment will make weapons grade material much more accessible.

Metz, W. D., "Laser Enrichment: Time Clarifies the Difficulty," *Science,* vol. 191, p. 1162 (1976).
  Identifies some important questions in development of laser enrichment techniques.

————, "Reprocessing Alternatives: The Options Multiply," *Science,* vol. 196, p. 284 (1977).
  Brief description of several possibilities for handling irradiated fuel.

NUREG-0116. "Environmental Survey of the Reprocessing and Waste Management Portions of the LWR Fuel Cycle" (supplement No. 1 to WASH-1248), U.S. NRC report NUREG-0116 (October 1976) (NTIS).
  Summarizes environmental impacts of LWR fuel reprocessing and waste management.

ORNL-4451. "Siting of Fuel Reprocessing Plants and Waste Management Facilities," Oak Ridge National Laboratory report ORNL-4451 (July 1971) (NTIS).
  Treats siting aspects, particularly those related to emissions, of reprocessing and waste management facilities.

Pigford, T. H., et al., "Fuel Cycles for Electric Power Generation," Teknekron report EEED 101 (January 1973, rev. March 1975); "Fuel Cycle for 1000-MW Uranium-Plutonium Fueled Water Reactor," Teknekron report EEED 104

(March 1975); "Fuel Cycle for 1000-MW High-Temperature Gas-Cooled Reactor," Teknekron report EEED 105 (March 1975). These are included in "Comprehensive Standards: The Power Generation Case," U.S. EPA report PB-259-876 (March 1975) (NTIS).

Estimates fuel cycle material flows for various reactor systems.

Pigford, T. H., and Ang, K. P., "The Plutonium Fuel Cycles," *Health Physics,* vol. 29, p. 451 (1975).

Estimates amount of material in fuel cycles that utilize plutonium as a recycle material.

Rochlin, G. I., "Nuclear Waste Disposal: Two Social Criteria," *Science,* vol. 195, p. 23 (1977).

Considers resistance to recovery or release and number of sites as criteria for selecting methods of waste disposal.

WASH-1248. "Environmental Survey of the Uranium Fuel Cycle," U.S. AEC report WASH-1248 (April 1974) (NTIS).

Summarizes environmental impacts of the LWR fuel cycle.

Zare, R. N., "Laser Separation of Isotopes," *Scientific American,* vol. 236, p. 86 (February 1977).

Describes how laser isotope separation works and how it might be used.

CHAPTER TWELVE

The Weapons Connection

THE SAME MATERIALS that sustain a barely critical chain reaction in a nuclear reactor can support a rapidly increasing reaction rate in other circumstances. Such an explosive buildup of the fission rate is a basic ingredient of nuclear weapons based on fission, whether as the main energy source, as enhancement, or only as the trigger. The other important requirement is that the fissile material be kept together long enough for a significant portion of the material to actually undergo fission. Construction of a fission bomb, therefore, requires assembling a suitable amount of fissile material in a way that yields a neutron multiplication factor (see Chapter 1 and Appendix E) well above 1, and keeping it assembled so that it doesn't "fizzle."

Nuclear weapons materials, in the weak sense, exist in any commercial nuclear power system. The principal fissile isotopes, ^{235}U, ^{239}Pu, and ^{233}U are the basic weapons materials. However, as used commercially, these materials have usually not been present in the form required for weapons. These devices require such a high density of fissile nuclei that the low enrichment fuels that are predominant in the commercial market are not suitable. The weapons grade material used in nuclear weapons is typically greater than 90% fissile, although lower grades can be used. The lowest fissile content practical for making a weapon is in the range 10% to 20%, much higher than the fissile content of fresh fuel for water-cooled reactors. Producing weapons grade material from this fuel takes considerable effort.

Before proceeding, we should note the amounts of material required for a nuclear weapon. This will depend on a number of factors, the details of which are not available to the public. Information that is relatively well-known is the "critical mass" associated with various weapons materials. Basically, this is the amount of material of specified composition and density which, when assembled into a spherical shape, will yield a (prompt) critical state; the injection of some neutrons is all that is necessary to start the chain reaction on its way. Ordinarily, in an actual weapon, the fissile material would be surrounded by a material, such as beryllium,

186

that is particularly good at reflecting escaping neutrons back into the fissile material. The quality of the reflector and the manner in which the critical mass is assembled strongly affect the amount of material required. For example, implosion, i.e., use of TNT to explosively compress the material, can substantially decrease the amount needed by sharply increasing the density of the material. However, to indicate an intermediate quantity, we simply assume material of normal density, but surrounded by a reflector.

In these circumstances, a critical mass of fissile *metal* is about 4 or 5 kg for ^{239}Pu and ^{233}U, and 11 kg for ^{235}U.[1] Use of an *oxide* reduces the density of fissile atoms, increasing the amount required by about 50%. If the material is not the isotope mentioned, but is diluted, the fissile amounts go up, but not very rapidly. Plutonium from a light-water reactor, for example, is about 20% nonfissile. Even so, about 8 kg of LWR plutonium constitutes a critical mass. However, as the fissile content gets very low, approaching 10%, the amounts required increase rapidly.

These are not large amounts of material; a 4-kg sphere of plutonium is smaller than a baseball. The real question is whether these materials can be obtained in a workable form or can easily be transformed to such a form. The pure fissile materials can be handled at close range if appropriate precautions are taken. How easy is it for nations, groups, or even individuals to obtain or produce these materials, in particular from commercial nuclear power? The answer to this question is not known. Nor is it known how best to prevent or discourage manufacture of weapons, whether from commercial nuclear materials or otherwise. It is nevertheless useful to indicate features of commercial nuclear power that are relevant to this question and to note some (though by no means all) of the nuclear power strategies that might discourage diversion of nuclear materials to weapons purposes. These strategies must be evaluated giving proper consideration to the desire of many nations to achieve improved uranium utilization, a goal that has usually been thought to require that fuel be reprocessed, as discussed in Chapter 10. Advanced reactor systems are described in more detail in Chapters 13 and 14.

NATIONS, GANGSTERS, AND NUCLEAR PROLIFERATION

The nuclear fuel cycle as it has been envisioned (Figure 11-2) would be dispersed in a way that, in principle, gives both nations and individuals legal or extralegal access to nuclear materials. If a nation is in possession of all types of fuel cycle facilities, it clearly has at its disposal both the materials and facilities with which to produce weapons grade material. In fact, a nation will be in this position if it has either an enrichment plant or a reprocessing plant. The first can convert low-enrichment uranium, the basic material of nuclear power, into highly enriched uranium, one of the basic materials for nuclear weapons. For nations that do not have enrichment facilities, a fuel reprocessing facility will serve, since this facility

[1] The suitability of specific materials depends on the type and sophistication of the device being built. Some isotopic mixtures may be particularly difficult to use, but this is not an easy matter to judge; for present purposes, we do not discriminate on the basis of this question.

processes spent fuel to extract plutonium, an adequate weapons material (although it may not be the easiest to employ). In the absence of either of these commercial facilities, a nation could construct a smaller facility, presumably a chemical extraction plant, devoted to extraction of weapons materials.

On the other hand, individuals, whether terrorists or gangsters, do not as a matter of course have access to nuclear materials. Theft is the means by which they might secure fissile material, and, unless their resources are substantial, the thieves would have to choose material that required a minimum of processing. For the LWR or LMFBR fuel cycle, this can be understood to mean stealing plutonium nitrate or plutonium oxide, either in pure form or mixed with uranium, possibly already incorporated in fresh (as opposed to spent) fuel. The alternative is for the group to have some sort of irradiated fuel reprocessing capability, not on the scale of a commercial plant, but still a substantial establishment because of the several processes that must take place behind heavy shielding.

Even presuming a nation or group has materials in hand, successful fabrication of a weapon cannot be presumed. Producing a nuclear weapon requires a determined effort by technically trained individuals with access to the proper materials and facilities. However, considering the general availability of basic informa-

Figure 12-1. DIVERSION SENSITIVITY OF LOW-ENRICHMENT URANIUM-PLUTONIUM FUEL CYCLES.
The figure is basically an LWR fuel cycle (cf. Figure 11-2). The sensitive facilities are the enrichment plant, and the fuel reprocessing and recycle fuel fabrication plants. The sensitive material pathways are those with extracted or freshly fabricated plutonium. (The fuel system for the LMFBR would be similar except that the uranium fuel need not be enriched and may, in fact, be enrichment plant tails.) The dashed line indicates uranium material used in plutonium-bearing fuel assemblies.

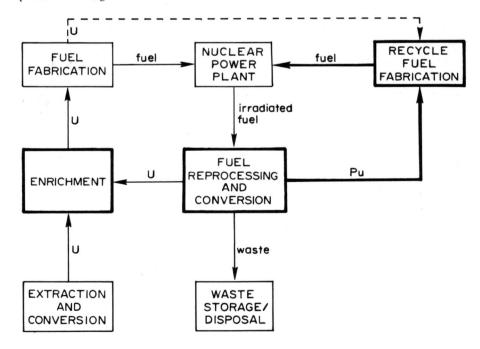

tion on weapons, it is prudent to assume that anyone who has weapons grade material has the capability of using it. For this reason, international and national entities take measures to make acquisition of weapons materials difficult, either diplomatically or physically. The effectiveness of these measures is the question.

The agency with responsibilities on an international scale is the International Atomic Energy Agency (IAEA). Although the primary purpose of the IAEA is to promote the peaceful use of nuclear energy, it undertakes, as an adjunct to this purpose, the inspection of national nuclear facilities. Signatories to the Nuclear Non-Proliferation Treaty submit to this inspection. This includes monitoring inventories of fuel to reveal any diversion. (It is interesting to note that the CANDU reactor can present special difficulties in this regard because fuel bundles are removed from the reactor on a daily, rather than yearly, basis.) However, should IAEA surveillance reveal diversion, it has no sanction at its disposal other than to invoke the disapproval of the international community. Moreover, many nations do not subscribe to the treaty. In fact, several nations are avowedly seeking to build nuclear weapons. As of 1977, India was the only nation known to have built a nuclear "device" from plutonium produced by a nonmilitary reactor. Other nuclear weapons states, the United States, Russia, England, France, and China, have used materials produced by expressly military facilities.

The difficulty of preventing diversion such as that of India will become more and more difficult as water-cooled reactor power plants spread, making plutonium-bearing spent fuel accessible to increasing numbers of nations; as fuel reprocessing technology is obtained by nations other than those that already have nuclear weapons, making recycle of plutonium an integral feature of their power generating systems; and as the LMFBR is introduced to more nations. This last development would vastly increase the amount of plutonium circulating. The exact amounts are unimportant, but on the order of 1000 kilograms of plutonium would be recycled per reactor per year. Excess bred plutonium from only one LMFBR would be over 100 kg/year, enough for 10 weapons (per reactor per year). Figure 12-1 indicates the flow of nuclear material for a low-enrichment uranium-plutonium fuel cycle; Figure 12-2 does the same for a thorium-uranium fuel cycle that requires highly enriched uranium as a feed material.

Existence of fuel reprocessing and of a large plutonium currency would also increase the accessibility of plutonium to individuals. The possibilities range from quiet milking of plutonium streams at a reprocessing plant or elsewhere to highjacking or robbing the plutonium bank. Preventing surreptitious theft requires a high level of employee surveillance and materials accounting. Even under these circumstances, plutonium embezzlement may be possible, except in unusually secure situations. Preventing blatant bank robbing may require highly secure (and armed) installations. A particularly vulnerable portion of the fuel cycle is transportation. Considering either employee plutonium snatching or armed assaults on fuel cycle installations, preventing determined and informed thieves may require extraordinary measures.

Diversion by nations or gangsters occurs in different circumstances. However, technical alterations may be possible that can substantially decrease the availability

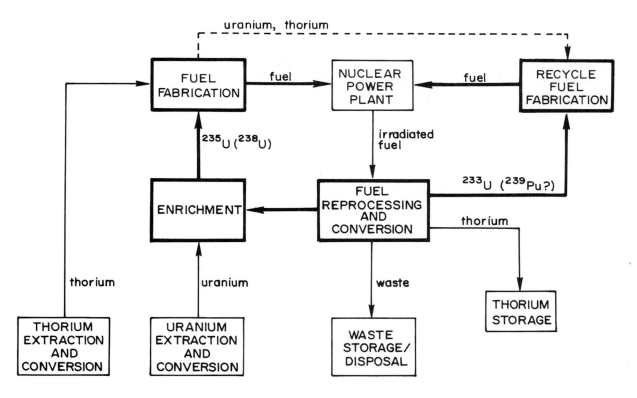

Figure 12-2. DIVERSION SENSITIVITY OF THORIUM-URANIUM SYSTEMS USING HIGH-ENRICHED ^{235}U FEED.

The flows for a standards HTGR or a thorium-cycle CANDU would be similar to those shown in the figure, although the details may differ. The enrichment, fuel reprocessing, and fabrication facilities process uranium (or plutonium) with high fissile content. The pathways connecting these facilities with each other and with the power plant carry sensitive material. The dashed line indicates ^{235}U, ^{238}U, or ^{232}Th which may be utilized at the shielded fabrication facility.

of nuclear materials or increase the effort to convert them to weapons use. Succeeding sections discuss some (in fact, only a small portion) of the possibilities now under consideration. They are expressly designed to make national diversion difficult, but the same measures will make it even harder for the individual thief.

For practical purposes, the alternatives may be assigned to two classes: those that do not reprocess spent fuel, and those that design the fuel cycle to maximize the effort required by a nation to obtain weapons materials. It will be obvious that the first class of alternatives is not suitable for a world or nation where nuclear power is rapidly expanding beyond the level supportable by natural uranium resources. The second class of alternatives permits rapid growth, but at the cost of more complex fuel cycle arrangements. In either case, because the array of reactors and fuel systems that might be considered to be more proliferation resistant than an LWR/LMFBR system is truly vast, we mention only a few of them, noting the principal factors to be considered in any comparison.

ONCE-THROUGH FUEL CYCLES

Water-cooled reactors, LWRs and CANDUs, are now operating on a "once-through" fuel cycle in practice. For CANDUs, this is intended, but for LWRs this state of affairs, if it continues, would represent a departure from previous intentions. An intentional once-through fuel cycle would include arrangements for long-term storage or final disposal of spent fuel, arrangements that have been thought out to some extent by the makers of CANDU. The question is whether such a once-through cycle, where the fuel is passed once through the reactor and then disposed of (see Figure 12-3), is practical in a world with limited uranium resources. And more to the point in the present context, what are the nonproliferation advantages?

Such an advantage may be based on the fact that, in a once-through fuel cycle, all fissile material remains in a form that is not directly usable for nuclear weapons. The plutonium emerging from a uranium-fueled reactor remains embedded in highly radioactive spent fuel; its extraction requires an explicit commitment by a nation and a substantial commitment by a terrorist group. A nation would have to build a specifically military reprocessing facility. A terrorist group would have a difficult time doing the same. For this reason, a once-through fuel

Figure 12-3. ONCE-THROUGH FUEL CYCLES.
If fuel is not reprocessed, highly enriched fissile material is not available, except to those who have an enrichment facility or are willing to reprocess the irradiated fuel independently of commercial facilities. In a once-through system, irradiated fuel is directly disposed of after minimal packaging and handling.

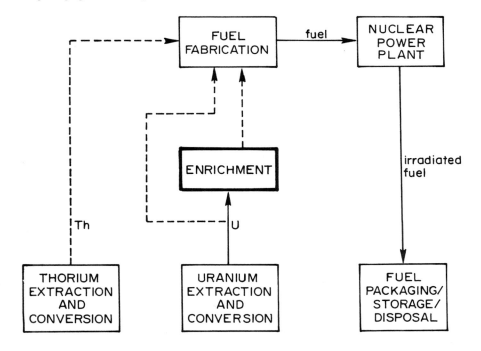

cycle is relatively resistant to diversion. Effective inspection procedures could sound a timely alarm, even for national diversion.

The difficulty is that such systems are not efficient in their use of uranium resources. A typical LWR that requires 4100 tons of U_3O_8 during its lifetime with recycle requires about 6000 tons without. A CANDU is notably superior, requiring less than 5000 tons with no recycle.

It is difficult to design once-through systems that have fuel-utilization much superior to the CANDU. Minor improvements in LWR operation can reduce resource requirements slightly. This would include reoptimizing the fuel design and management for a once-through, rather than a closed, fuel cycle. More detailed fuel management, simply responding to increased resource values, would also slightly reduce uranium requirements. Overall, such changes may reduce requirements by about 10%, but they cannot bring the LWR to match the CANDU, which is already designed for once-through operation, has an intrinsic capability for detailed and continuous fuel management, and has a superior moderator/coolant (but may pose a difficult inspection problem).

A more serious alteration of the LWR is to introduce some heavy water into the system, particularly just after refueling, when the reactor would ordinarily use

Figure 12-4. INTERNATIONAL FUEL REPROCESSING (AND FABRICATION) CENTER. International control of sensitive fuel cycle facilities may keep sensitive materials out of national hands. In addition, the lines of communication are shortened, reducing vulnerability to terrorist action.

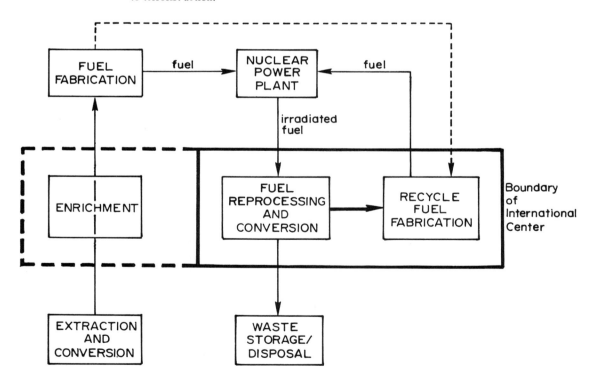

neutron absorbers to compensate for excess fissile material. The neutrons are under-moderated, so that a greater portion are absorbed by ^{238}U. In such a "spectral shift controlled reactor" (SSCR), therefore, neutrons that would have been lost to neutron absorbers convert ^{238}U to ^{239}Pu, which is then available as additional fissile material for extending the fuel life and reducing the demand on natural fissile resources. The uranium utilization of an SSCR would be in the vicinity of an ordinary (once-through) CANDU or an LWR with uranium and plutonium recycle. It would be significantly better than the once-through LWR.

An alternative to disposing of spent LWR fuel is to use it, in particular, in a reactor designed for very low enrichment, such as a CANDU. Substantial technical problems are associated with this option. To avoid reprocessing, either the LWR fuel or the CANDU fuel channel must be redesigned. The fuel requirements for a tandem LWR/HWR system would be comparable to the best that a once-through CANDU or an LWR with recycle could do.

Finally, even the usual high-temperature gas-cooled reactor, with its highly enriched (93%) uranium and ^{233}U recycle, can be adapted to a once-through fuel cycle. Using low enrichment uranium (about 10% ^{235}U) on a once-through cycle, the HTGR is considerably superior to an LWR without recycle, and comparable to a CANDU, as far as uranium utilization is concerned.

A brief scan of these possibilities shows that, although systems can have superior fuel utilization characteristics to an LWR, the amount of improvement is limited. In fact, the systems noted, the CANDU, the SSCR, the LWR/HWR tandem, and the low enrichment HTGR all have lifetime uranium requirements in the range of 4000 to 5000 tons of U_3O_8. Some improvement on this is possible, but substantial improvement requires recycle.

INTERNATIONAL CENTERS AND DENATURING

Reactor systems that depend on fuel recycle now hold the eye of the international community. Although other nations have shown some concern over the possibility of proliferation, many are indifferent and few appear as willing as the United States to consider seriously whether fuel recycle should be given up in the interest of international stability. One reason for this is that few nations have as diverse an energy supply as the United States. Regardless of the reasons, however, the very fact that few nations appear at all willing to forego fuel reprocessing makes it useful to examine how recycle (as opposed to once-through systems) may be made relatively proliferation resistant. The two general approaches discussed below are the internationalization of fuel cycle facilities and the modification of fuel materials so that they are not easily fabricated into weapons.

The basic idea of an international center is to extract, manipulate, or utilize sensitive materials under strict international control, as suggested in Figure 12-4. This would not of itself be a panacea because we are left with the difficulty of careful material monitoring within the center to prevent theft and with the possibility that nations (or groups) will make the commitment to carry out surreptitious fuel processing outside the center. Strictly speaking, this center would also include any necessary enrichment capacity, even though this part of the problem is often left out, perhaps because the advanced system usually considered, the LMFBR, does not require enrichment facilities. However, as we have noted, thorium-uranium reactors often require enriched uranium (even highly enriched) as feed to the fuel cycle.

Restriction of fuel processing to international centers keeps relatively pure fissile material out of national hands and eliminates movement of such materials along open transportation links. The pure material may still exist at various points inside the center. It is possible to reduce even this vulnerability by designing the process steps so that sensitive materials remain diluted or contaminated with other materials to such an extent that they are dangerous or cumbersome to steal. In fact, one form of this idea has appeared in connection with the possibility of plutonium recycle for LWRs in the United States. The plutonium, instead of being extracted (from LWR fuel) in a relatively pure form, could be co-extracted with uranium in a mix that could then be fabricated directly as LWR-mixed oxide fuel. In this way, a thief would have to carry off and process much more material than if the plutonium were converted to pure plutonium oxide.

It is important to note, too, that the fuel cycle shown in Figure 12-4 gives no indication of the form of the recycle fuel or of its sensitivity to diversion. Mixed oxide fuel, whether for an LWR or an LMFBR, contains plutonium that could be used for weapons if chemically extracted. Because this mixed oxide fuel is not very radioactive (it is just plutonium and uranium, uncontaminated by fission products), extracting the plutonium is relatively easy, possibly with close handling and no heavy shielding. The only barrier to a terrorist is the difficulty of stealing the material and, perhaps, of processing a large amount of material; but with LMFBR fuel (containing 10% to 20% plutonium), even the amounts are not very large. In any case, neither of these difficulties is an effective barrier to diversion by a nation. A much more effective barrier is to release no fresh fuel that can be converted to weapons materials by easy chemical procedures.

This suggests a second class of measures, intended to make it difficult to convert recycle fuel to weapons. These measures depend on diluting the fissile material either isotopically or radioactively. The first type of dilution is possible with only one of the fissile materials bred in a reactor. The most common material, plutonium, is weapons material and there is no suitable way of diluting it isotopically, i.e., as ^{235}U is naturally diluted by ^{238}U. The alternative bred material, ^{233}U, can be diluted (or "denatured") by ^{238}U. If the fissile content does not exceed 10% or so, conversion to weapons material requires an enrichment facility, even more of a commitment (at the present time) than reprocessing. Since the thorium-uranium cycle produces ^{233}U, this cycle appears adaptable to a more proliferation-resistant form than the uranium-plutonium cycle. However, as noted below, denaturing with ^{238}U necessarily leads to production of plutonium, which must be used or disposed.

The second type of mixing, with radioactive materials, is intended to make the recycle fuel hot to handle, so that it would require shielded chemical processing before being fabricated into a weapon. The radioactivity would also make it more difficult to steal the fissile material or the assemblies containing it. The fissile mixture can be left highly radioactive from the time of its extraction by leaving fission products in it. Alternatively, the extracted fissile, whether ^{239}Pu or ^{233}U, can later be "spiked" with some particular radioactive isotope(s). In general, spiking leads to a somewhat simpler fuel cycle than denaturing because it does not influence, in any intrinsic manner, the flow of fissile and fertile material.

Denaturing ^{233}U with ^{238}U, on the other hand, can alter the fuel cycle considerably. A pure thorium-uranium fuel cycle produces only uranium (particularly ^{233}U) with no plutonium. The system could be started on highly enriched uranium, or even on plutonium. The principal recycle fuel is thorium/^{233}U. Denaturing the ^{233}U adds fertile ^{238}U, which produces plutonium. How much? As a rough comparison with 3% enrichment uranium, consider a ^{233}U–^{238}U–^{232}Th mixture in a ratio of 12:88:300, so that ^{233}U is 3% of the mixture. (In an actual reactor, about 4% would be needed, because the cross-sections differ from those in a ^{235}U–^{238}U system.) Ignoring cross-section differences, one would expect about 23% as much plutonium to be produced as would be with the ordinary 3% ^{235}U–97% ^{238}U fuel. In fact, because of differences in cross-section, the amount of plutonium produced would be smaller, but it is still a substantial amount.

Disposal of this plutonium as waste would considerably reduce the system conversion or breeding ratio. Alternatively the system's proliferation resistance could still be maintained by using the plutonium within the international center. It could be refabricated and "burned," in either a thermal or fast reactor. However, the principal fertile fuel in either case would be thorium, so that the reactor consumes plutonium and produces ^{233}U. Low enrichment uranium fuel from the center is sent to dispersed reactors, which return ^{233}U fuel bearing both plutonium and ^{233}U. This basic concept is suggested in Figure 12-5. However, there are many variations, depending on whether the power plants are high-conversion thermal reactors or breeders. The most effective use of any bred plutonium is as feed to a breeder in the center, and perhaps also outside. If plutonium is used outside the center, radioactive spiking would reduce the proliferation potential.

In a situation where the nuclear system is growing slowly, such as the United States, use of high-conversion thermal reactors seems quite feasible. However, a system that is growing rapidly to a large scale may require a fast breeder to avoid exceeding resource limitations. Although such a device can operate on a thorium-uranium fuel cycle, denaturing the ^{233}U still produces much plutonium which must be utilized to maintain a breeding ratio over 1. The ease of diversion can, of course, be reduced with spiking. But it does appear that a rapidly growing nuclear economy will either have relatively pure ^{233}U as its currency or low enrichment ^{233}U plus plutonium.

ALTERNATIVE ROUTES TO NUCLEAR WEAPONS

The most difficult aspects of the proliferation question are probably unrelated to commercial nuclear power and its use of plutonium. One imponderable is the likelihood that a nation having nuclear weapons will use them; more broadly

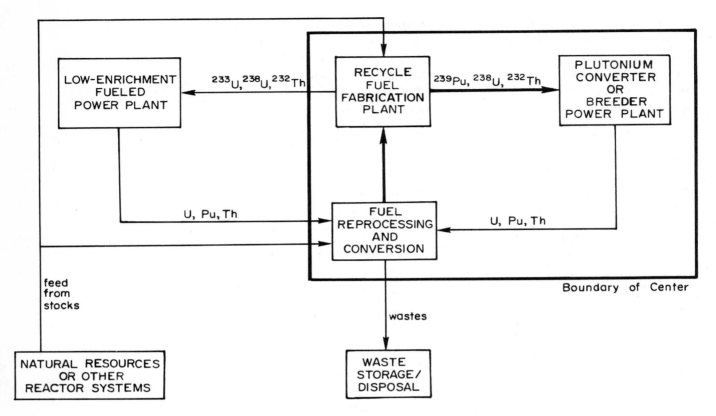

Figure 12-5. INTERNATIONAL FUEL REPROCESSING AND PLUTONIUM UTILIZATION CENTER.
The site of an international fuel reprocessing and fabrication center may also be used for "burning" plutonium from power plants that are off site. The "burner" could be either a thermal or fast reactor, which is designed to accept plutonium from other reactors and which, preferably, produces ^{233}U, which may be supplied in denatured fuel to other reactors.

put, how does the spread of nuclear weapons affect world stability (and vice versa)? It is ironic that one of the nations most actively seeking to prevent the spread of weapons, the United States, is the only nation to have used one in war. The fact that none has been used in three decades demonstrates that ownership does not necessarily imply use.

A difficulty more directly connected with the present discussion is that ownership of commercial nuclear facilities and construction of weapons may not be connected. It certainly is true that many of the major developers of nuclear reactors have nuclear weapons. But none of the major nuclear powers reached that status via commercial nuclear development. Rather, the converse has been true: commercial developments have often rested on previous military work. The notable exception to this pattern is India (and possibly some nations that have not gone "public"), who fabricated a weapon from fuel from a civilian reactor.

A serious problem is that other ways to create weapons exist, even if diversion of commercial materials is made technically difficult. These include construction of plutonium production reactors or enrichment facilities, the two routes used by the major nuclear powers. The first route may seem substantial, but it can be largely circumvented by use of the small research reactors now in the possession of many countries. Since one or a few nuclear weapons do not require much plutonium, a large power reactor is not necessary. What is required is the ability to reprocess irradiated fuel.

The spread of enrichment facilities would put highly enriched uranium, probably the easiest material to make a bomb of, within the grasp of other nations. Present facilities are very large, expensive, and difficult to fabricate. A major question is the extent to which development of an effective method of laser enrichment would encourage other nations to attempt their own enrichment. Laser methods would certainly use a high form of technology, but the size and expense of the facilities would probably be reduced. Laser isotope separation could, of course, simply lead to more efficient commercial nuclear operation. However, it may open one more way to nuclear weapons.

As a rule, though, these approaches to weapons production require a substantial commitment to that purpose, if not in economic terms, at least in terms of making a conscious decision and devoting a significant amount of skilled effort over a long period of time. A commercial nuclear system that circulates large amounts of clean plutonium would circumvent these difficulties, as would the national control of a fuel processing facility. An international commitment to limiting the uncontrolled availability of these facilities and to avoiding the circulation of fissile material suitable for weapons would certainly lengthen the time and effort required to acquire nuclear weapons. It would at least decrease the likelihood that the presence of peaceful nuclear facilities will lead directly to the production of warlike devices.

Bibliography — Chapter Twelve

"NTIS" indicates report is available from: National Technical Information Service, U.S. Department of Commerce, 5285 Port Royal Road, Springfield, VA 22161.

APS-1977. L. C. Hebel et al., "Report to the American Physical Society by the Study Group on Nuclear Fuel Cycles and Waste Management," (July 1977). To be published as supplement to *Reviews of Modern Physics,* vol. 50.
  A study of nuclear fuel cycles that considers their relationship to the possibility of diversion of nuclear materials.

The Bulletin of the Atomic Scientists, a magazine that has long had an interest in the problem of nuclear proliferation, in its October 1977 issue has several articles that pertain to this problem: J. S. Nye, "Time to Plan for the Next Generation of Nuclear Technology," p. 38; S. Eklund, "We Must Move Forward with All Deliberate Speed," p. 42; and B. Stewart, "Some Nuclear Explosions Will Be Necessary," p. 51.

Casper, B. M., "Laser Enrichment: A New Path to Proliferation," *Bulletin of the Atomic Scientists*, vol. 33, p. 28 (January 1977).
  Argues that work on laser enrichment of uranium should be suspended until its implications are understood.

EPRI NP-359. "Assessment of Thorium Fuel Cycles in Pressurized Water Reactors," Electric Power Research Institute report EPRI NP-359 (February 1977) (NTIS).

Evaluates the fuel requirements for PWRs operated on thorium cycles.

Feiveson, H. A., and Taylor, T. B., "Security Implications of Alternative Fission Futures," *Bulletin of the Atomic Scientists*, vol. 32, p. 14 (December 1976).

Discusses how the type of nuclear fuel cycle may influence the ease with which nations can divert commercial materials to weapons.

Ford Foundation/MITRE Corporation, "Nuclear Power: Issues and Choices," Report of the Nuclear Energy Policy Study Group (Ballinger, Cambridge, Mass., 1977).

Considers future direction for nuclear power, emphasizing the influence of choices on the problem of nuclear proliferation.

Foster, J. S., and Critoph, E., "The Status of the Canadian Nuclear Power Program and Possible Future Strategies," *Annals of Nuclear Energy*, vol. 2, p. 689 (1975).

Summarizes the current Canadian program and fuel cycles which are of interest for the future.

Gall, N., "Atoms for Brazil, Danger for All," *Bulletin of the Atomic Scientists*, vol. 32, p. 4 (June 1976).

Examines the implications of the sale of fuel cycle facilities to Brazil.

Krass, A. S., "Laser Enrichment of Uranium: The Proliferation Connection," *Science*, vol 196, p. 721 (1977).

Examines how seriously laser enrichment might exacerbate the problem of nuclear proliferation.

Merrill, M. H., "Use of the Low Enriched Uranium Cycle in the HTGR," General Atomic Company report GA-A14340 (March 1977).

Compares low-enrichment once-through uranium fuel cycles for the HTGR with other HTGR and PWR alternatives.

Meyer, W., Loyalka, S. K., Nelson, W. E., and Williams, R. N., "The Homemade Nuclear Bomb Syndrome," *Nuclear Safety*, vol. 18, p. 427 (1977).

Argues that it is not easy to steal nuclear materials and fabricate a nuclear weapon.

"Nonproliferation Alternative Systems Assessment Program: Preliminary Program Plan" (draft), U.S. ERDA report (May 1977).

Outlines the ERDA program for assessing the proliferation potential of alternative nuclear systems.

"Nuclear Proliferation and Safeguards," Congress of the United States Office of Technology Assessment (Praeger, New York, 1977).

Examines the problem of nuclear weapons proliferation and the impact that the spread of commercial nuclear facilities may have.

NUREG-0001. "Nuclear Energy Center Site Survey — 1975," 5 vols., U.S. NRC report NUREG-0001 (January 1976) (NTIS).

Surveys sites in the United States that are suitable for multiple nuclear facilities.

Rotblat, J., "Controlling Weapons-Grade Fissile Material," *Bulletin of the Atomic Scientists*, vol. 33, p. 37 (June 1977).

Discusses safeguards to prevent diversion of nuclear materials by nations or criminals/terrorists.

Seghal, B. R., Lin, C. L., and Naser, J., "Performance of Various Thorium Fuel Cycles in LMFBRs," Electric Power Research Institute, *EPRI Journal*, vol. 2, p. 40 (September 1977).

Compares the amounts of thorium, uranium, and plutonium that would be supplied to and removed from LMFBRs operated on various thorium cycles.

Willrich, M., and Taylor, T. B., "Nuclear Theft: Risks and Safeguards," (Ballinger, Cambridge, Mass., 1974).

Analysis of the possibility of weapons being fashioned from materials stolen from the commercial nuclear industry.

Wilson, R., "How to Have Nuclear Power Without Weapons Proliferation," *Bulletin of the Atomic Scientists*, vol. 33, p. 39 (November 1977).

Argues that there are more effective ways to limit the spread of nuclear weapons than by withholding commercial nuclear power from other nations.

Part IV

Advanced Reactor Systems: Breeders, Near-Breeders, and What-Not

Second-generation nuclear power can utilize a large array of systems, some similar to the water-cooled reactors of the first generation, and some not. The choice of systems will depend on several considerations, many of which were outlined in Part III. The fundamental purpose of advanced systems has been to improve our utilization of uranium and thorium resources. But other factors — such as possible aversion to a plutonium economy, whether because of plutonium's biological toxicity or because of its proliferative tendencies — may turn out to have great influence.

The purpose of Part IV is to describe many of the important advanced nuclear systems. It will be obvious that the primary consideration in their design is to improve resource utilization. The most resource-efficient devices (at least in a fast-growing nuclear economy) are the fast breeder reactors. These are described in Chapter 13, which emphasizes the liquid metal fast breeder reactor, the central feature of most countries' advanced reactor programs. Chapter 14 describes advanced thermal reactors, particularly those that utilize a thorium-uranium fuel cycle. The last chapter, 15, introduces some systems that are basically different from pure fission reactor systems in that at least one of their nuclear components is not an ordinary fission reactor. Of the systems described in Part IV, the only one to have reached the demonstration phase is the liquid metal fast breeder.

CHAPTER THIRTEEN

Fast Breeder Reactors:
The LMFBR and Relatives

COMMERCIAL NUCLEAR SYSTEMS now employ thermal reactors operating on the uranium-plutonium fuel cycle (see Part II). Building a reactor that produces more fissile material than it consumes, or even comes close to this condition, requires a radical departure from this approach: as discussed in Chapter 12, the uranium-plutonium cycle is not neutronically capable of break-even in a thermal reactor.

Improving conversion to the point of break-even or beyond can be accomplished in one of two ways. The neutron spectrum can be kept energetic enough to take advantage of the fact that ^{239}Pu yields many more neutrons at high energies. Or ^{232}Th can be used as the fertile material, yielding ^{233}U, which has a relatively high neutron yield with both thermal and fast incident neutrons (see Appendix C).

We begin with the fast reactor systems, simply because they are more highly developed, having reached the demonstration phase in several countries. Taking care not to moderate the neutrons makes use of ^{239}Pu's high value of eta (the number of neutrons produced per neutron absorbed; see Table C-4) in a fast spectrum. For every neutron absorbed by ^{239}Pu, about 2.7 neutrons can be produced, easily supplying enough to continue the chain reaction and to convert an atom of ^{238}U to one of ^{239}Pu, replacing the one that was destroyed.

A difficulty arises because the fission cross-section (probability of fission) is so low at high energies that proportionately more fissile material must be contained in the fuel to maintain a chain reaction. Otherwise, simple captures (notably by ^{238}U) absorb enough neutrons to quench a chain reaction. For this reason, the core has relatively high fissile concentrations, in the vicinity of 15%, as compared with 3% for an LWR and 0.7% for a CANDU.

The basic nuclear configuration of all fast breeders is similar. In conventional terms, the reactor is divided into an active "core" region and a "blanket" of fertile material This core[1] is a relatively compact critical assembly, with about 15% fissile

[1] In other reactors, this fissile-bearing region may be referred to as the "seed" region (see LWBR, next chapter).

material and the remainder fertile. The core can be compact because no space need be taken up by the moderator. Neutrons leaking from the core are absorbed in the fertile blanket material. With a ^{239}Pu–^{238}U core and a ^{238}U blanket, breeding ratios above 1.2 appear to be possible; as noted in the next section, a more heterogeneous configuration (without the strict division into core and blanket) may yield a higher breeding ratio. It is also feasible to use the thorium-uranium cycle, particularly in the blanket region.

Two other characteristics of fast breeders are worthy of note. The first is that the use of a compact core directly implies a high power density, at least as compared with thermal reactors. This imposes a need for a coolant system with good heat-transfer properties. The second characteristic of fast breeders is that the fuel is typically oxide pellets with stainless steel (rather than Zircaloy) cladding. However, other fuel forms, including both carbides and metal, may be advantageous in more advanced designs.

LIQUID METAL FAST BREEDER REACTORS

The liquid metal fast breeder reactor has been the mainstay of every breeder development program, including that of the United States. As indicated by the name, the LMFBR uses a liquid metallic coolant, a suitable choice where good heat-transfer characteristics are required. The coolant of choice is sodium. The basic heat transfer scheme for an LMFBR is shown in Figure 13-1. As indicated there, and discussed below, the turbogenerators are still driven by steam.

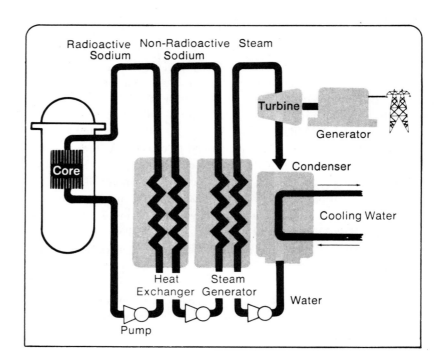

Figure 13-1. SCHEMATIC LIQUID-METAL FAST BREEDER REACTOR POWER PLANT.
In an LMFBR, the core is cooled by liquid sodium. A heat exchanger transfers heat from this coolant to sodium in an intermediate loop. The sodium in this loop raises steam in a steam generator. This steam then drives the turbogenerator. (Figure reproduced from ERDA-76-107.)

The core is a compact arrangement of fuel assemblies which are similar to those used in light-water reactors except that the fuel material is more highly enriched, the fuel pins typically have a smaller diameter, and the cladding is stainless steel rather Zircaloy. A possible form for the fuel assembly is shown in Figure 13-2.

Instead of using ^{235}U or ^{233}U as the fissile material, the present choice is to load the core with ^{239}Pu; this is the inevitable product of the current generation of light-water reactors, and, because of its high value of eta (neutrons produced per neutron absorbed) in a fast spectrum, it is much more profitably used in the breeder than in thermal reactors. (On the other hand, ^{235}U probably does not yield enough neutrons for a practical breeder; this is not true of ^{233}U, but the latter nuclide is not currently being produced in any large quantity.) The fertile nuclide now chosen is ^{238}U.[2] This is expected to produce more than enough ^{239}Pu to replace that which is consumed in the core. The "core" itself would be a mixture of oxides of plutonium and uranium. Surrounding this core would be a blanket of uranium. In both cases, the uranium would be almost entirely ^{238}U. Fission would occur primarily in the core region and conversion would occur in both regions. Regular reprocessing would be needed to recover the bred fissile material.

The blanket assemblies would have rods with the same composition throughout. Core assemblies, on the other hand, would have fissile loads along the portion of the rods that constitutes the core, but only fertile material at the top and bottom of the rods. In this way, the entire core is surrounded with blanket regions. An alternative to this clear separation of "core" and "blanket" is a heterogeneous design in which the core is subdivided into smaller regions, each consisting of a mixture of assemblies with fissile and fertile loading.

Power generation in the regions with heavy fissile loading will be quite intense compared with thermal reactors. As a result, the coolant (now a minimal moderator) must have good heat-transfer properties. In the present case, the LMFBR, this has led to the choice of sodium as the coolant. A metal that is liquid over a large range of temperature, sodium can successfully cool the very compact core, with its high power density. Additional advantages of importance are that it can be used at essentially atmospheric pressure, thereby making design easier, and that it can operate at a high enough temperature to permit a higher plant efficiency (see Appendix D) than water-cooled reactors can achieve. Sodium, however, is highly reactive chemically. At reactor operating temperatures, it burns if exposed to air; moreover, it reacts violently with water. Stringent efforts must be made to prevent any breaks or corrosion that lead to sodium leakage.

In all LMFBR designs, the sodium that cools the core is not used to raise the steam that drives the turbogenerators. Instead, there is an intermediate sodium loop (see Figure 13-1) which avoids the possibility of releasing radioactive sodium during any steam generator problem. This requires use of an intermediate heat exchanger as an interface between the primary and secondary sodium loops. It has the effect of more effectively isolating the primary sodium, and thus the sodium-filled reac-

---

[2] However, as noted later, this is not necessary; thorium may be a useful blanket (and even core) material.

Figure 13-2. LIQUID-METAL FAST BREEDER FUEL ASSEMBLY AND ROD (ELEMENT).

Fuel rods and assemblies for an LMFBR are similar to those for LWRs. Important differences are that the cladding for an LMFBR is stainless steel. Furthermore, because the LMFBR core is divided into "core" and "blanket" regions, the fuel rods are not uniform. The rod and assembly shown are for the "core" region. The portions of the rod and assembly labeled "core" contain moderate enrichment fuel (conventionally, depleted uranium plus plutonium), and the regions labeled "blanket" contain fertile material (conventionally, depleted uranium). (Figure reproduced from ERDA-1535.)

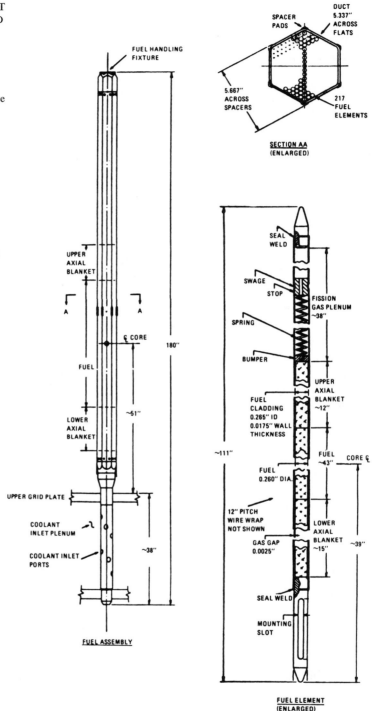

tor, from any water. It does not, however, eliminate the difficulty of designing steam generators which effectively separate sodium and water.

Two basic types of LMFBR are being considered. The form most favored by European countries is the "pool" type, in which not only the core, but also a number of other components are contained in the reactor vessel. (An example of this is given in Figure 13-3, which shows the French Phenix reactor.) The vessel is filled with sodium at roughly atmospheric pressure, in which are immersed the core, refueling machines, the primary coolant pump, and the intermediate heat exchanger, so that the entire primary sodium loop is contained in the same vessel. This assembly markedly reduces the amount of external piping. The alternative scheme, called the "loop" arrangement, has until recently been anticipated in the United States and is more similar to conventional LWR systems in that individual components of the heat transfer system are connected by pipes, and the reactor vessel contains only the core and associated equipment. Such a system would be employed in the proposed Clinch River demonstration breeder reactor, as shown schematically in Figure 13-4. (Figure 13-5 shows the type of reactor vessel typically employed in a loop system.) In either arrangement of the primary system, the vessels containing primary components are surrounded by guard vessels, so that any rupture of the primary system does not lead to a large loss of sodium. Finally, in either, the secondary sodium proceeds from the intermediate heat exchanger to a steam generator, which produces steam for driving the turbines.

In any LMFBR, an effort is made to minimize the amount of time the reactor is shut down for refueling. In either the pool or loop arrangement, use is often made of a rotating plug on the reactor vessel closure head. On this plug will be mounted an in-vessel fuel transfer machine, as well as the control rod drives, which are disengaged from the core when the plug rotates, enabling the transfer machine to move fuel from the core to an in-vessel holding point, or vice versa. In the pool arrangement, this holding point may amount to an actual storage drum, and fuel stored therein may be allowed to decay before it is exchanged (using an external machine) with fuel outside the vessel. This exchange may be accomplished while the reactor is still running. In the loop arrangement, decay storage must be outside the reactor vessel.

An important question for fast breeders is that of fuel lifetime. In thermal reactors, only a small percentage of the fuel atoms fission before the fuel is removed for storage or reprocessing. In fact, since the fissile load is typically less than 4% of the fuel, and conversion ratios are small, it is clear that the "burnup"[3] is only 2% or 3%. At burnups more than this, the fuel material and the cladding may suffer enough damage to prevent continued use of the fuel. (Moreover, in thermal reactors, one would wish to reprocess to remove neutron poisons.) But the breeder has a fissile loading of 15% or more, and the power density (and, correspondingly, the per volume rate at which nuclei fission) is much greater than in water-cooled reactors. For the fuel assemblies to have a sensibly long lifetime in the reactor, the

[3] Burnup may be given as the percentage of heavy metal fissioned (including all fuel materials, both fissile and fertile) or in terms of the amount of energy released per unit mass (in MWd/Te).

Figure 13-3. SCHEMATIC "POOL" TYPE LMFBR REACTOR VESSEL. In a "pool" arrangement, the entire primary system of the LMFBR is contained in the reactor vessel. Not only the core is immersed in the sodium coolant, but also the refueling machines, the primary pump, and the intermediate heat exchanger. The sodium that leaves the reactor vessel is that of the intermediate loop. The vessel shown is that of the French demonstration LMFBR, Phenix. (Figure courtesy of C. Pierre L. Zaleskie.)

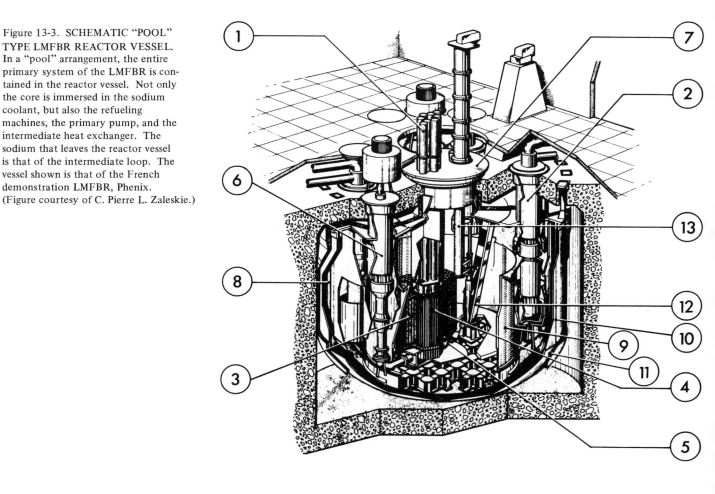

LEGEND

1. control rod drives
2. intermediate exchanger
3. lateral neutron shielding
4. core
5. fuel support slab
6. primary pump
7. rotating plug
8. main vessel
9. primary vessel
10. double envelope vessel
11. primary containment
12. transfer ramp
13. transfer arm

Net electric power	233 MWe
Thermal power	563 MWth
Date of criticality	4/1973
Fuel material	UO_2-PuO_2
Enrichment	
(volume fissile/total volume)	
Central core	19.2%
Outer core	27.1%
Blanket	Natural or depleted U
Cladding material	SS 316
Initial critical mass	735 kg [239] Pu equivalent
Maximum expected burnup	50,000 MWd/Te
Overall breeding ratio	1.16

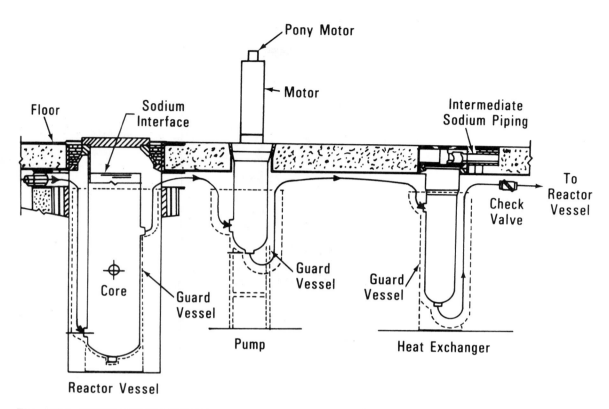

Figure 13-4. SCHEMATIC "LOOP" ARRANGEMENT FOR LMFBR PRIMARY SYSTEM. In a "loop" arrangement, the principal components of the LMFBR primary system are connected by piping. Much as in the case of LWRs, the reactor vessel, primary pump, and heat exchanger are separate; in this case, each is surrounded by a guard vessel to prevent large losses of sodium in the event of a break in the primary system. The figure shown is a schematic for the Clinch River demonstration breeder reactor. (Figure reproduced from ERDA-1535.)

burnup that they can tolerate must be higher, around 10%. The United States has 15% as the maximum burnup that fuel assemblies should stand, corresponding to a thermal energy generation of about 150,000 MWd/Te. The anticipated discharge burnup for LWRs is about 30,000 MWd/Te for the average assembly. (Average and maximum burnup may differ by up to a factor of two in a given reactor.) A particular consideration with such high burnup is that the stainless steel cladding may begin to swell and distort, thus damaging the fuel rods and even blocking the coolant flow. The relatively compact LMFBR core implies a greater neutron flux through the core structure than typically occurs in thermal reactors, giving rise to the possibility of swelling and alteration of the core configuration over a period of time. Restraints are typically included in the LMFBR core design in order to prevent such changes.

In general, the possibility of changes in the core configuration is of more concern in a fast reactor than in thermal reactors because the fuel in the former case has a substantially higher percentage of fissile material. Changes in core geome-

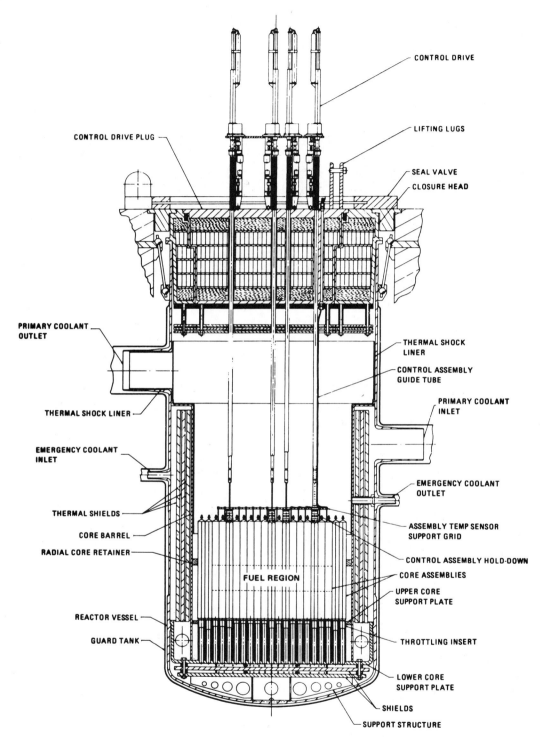

Figure 13-5. LMFBR REACTOR VESSEL FOR "LOOP" ARRANGEMENT.
The reactor vessel for a "loop" arrangement is generally similar to that for a pressurized-water reactor, although the details of the components are very different. The vessel shown is of the type that might be used in a commercial LMFBR in the United States. (Figure reproduced from ERDA-1535.)

try could more easily result in significant changes in the multiplication factor, leading to concern about the possibility of a core disruptive accident.[4] This concern is aggravated by the fact that development of a void in the coolant (due to, say, boiling) reduces the absorption and moderation of neutrons, both reductions leading to increases in the multiplication factor in a fast reactor. (In a thermal reactor, a void and the resulting decrease in moderation reduces the multiplication factor.) Care must therefore be taken that the design prevents significant changes in core geometry and that intrinsic neutronic feedback mechanisms are sufficient to prevent coolant voids, at least those that could propagate to affect large portions of the core. The latter mechanisms are basically intended to provide a decrease in the multiplication factor (or reactivity, see Appendix E) as the temperature rises. An important contributor to this negative temperature coefficient for fast reactors is the "Doppler" effect (discussed in Appendix E): as the temperature rises, the effectiveness of resonance capture (without fission) in removing neutrons from the system is increased, so that the fission rate is decreased. Finally, concern is often expressed that the prompt neutron lifetime (the time it takes for a given neutron to be absorbed, thereby producing the next generation neutron) is much shorter for fast reactors than for thermal; this time is about 5×10^{-7} sec, roughly one thousandth the comparable lifetime of a thermal-reactor neutron, which spends a relatively large amount of time slowing down and being absorbed. This would be of greatest concern if the multiplication factor exceeded 1 by so much that the reactor was critical on prompt, as opposed to delayed, neutrons alone. (See discussion of Appendix E.) Since this is a condition to be avoided in any type of reactor system, the short neutron lifetime does not, of itself, cause additional problems.

In general, the intrinsic safety aspects of fast reactors differ from those of thermal reactors. As indicated above, some of the differences work to the disadvantage of the designer under some conditions. However, one must note that other features, such as the good heat transfer properties of sodium and the fact that it can be utilized at low pressure and over a large temperature range, work to the designer's advantage. The brief discussion here cannot adequately describe the complex interplay of various factors important to reactor safety.

The British, French, and Russians already have prototype LMFBRs operating, each with outputs of 250 to 350 MWe; one of these was shown in Figure 13-3. The corresponding United States reactor, the Clinch River breeder reactor (CRBR), has until recently been scheduled for completion in the 1980s. This is being preceded by the "fast flux test facility" (FFTF), an important component of the breeder fuel testing program. After the Clinch River demonstration, larger commercial reactors would be built. However, substantial changes are being considered for the U.S. LMFBR program, including possible indefinite deferment of commercialization.

Some of the major design parameters for these three components of the U.S. program are shown in Table 13-1. Most plans call for commercial LMFBRs with thermal efficiencies of about 40%, breeding ratios of 1.2 or more, and doubling times of 10 to 20 years. The CRBR was intended to have a breeding ratio of 1.2

---

[4] In case a significant portion of the core should ever melt, the design must prevent the high-fissile-load fuel from reassembling (for example, at the bottom of the reactor vessel) into a critical configuration.

TABLE 13-1.
Comparison of Liquid Metal Fast Breeder Reactor Design Parameters

Design Parameters	FFTF	CRBR	Commercial and Commercial Prototype
Overall Plant			
Thermal power (MWth)	400	975	3,800
Net electric power (MWe)	NA	350	1,500
Overall plant efficiency (percent)	NA	35.9	39.5
Net plant heat rate (Btu/kWh)	NA	9,507	8,650
Plant capacity factor	NA	0.75	0.85
Number of primary loops	3	3	3
Containment diameter/depth below operating floor/height above operating floor (ft)	135/80/109	186/85/172	168/86/163
Reactor			
Fuel material	Oxide	Oxide	Oxide
Cladding material	SS 316	SS 316	Low swelling/SS 316
Fuel rod diameter (in.)	0.23	0.23	0.27
Fuel rod pitch/diameter ratio	1.26	1.26	1.24
Number fuel rods/assembly	217	217	271
Number core assemblies	76	198	318
Number blanket assemblies	None	150	234
Number control/safety assemblies	6/3	15/4	27/4
Core height/diameter (ft)	3.0/4.0	3.0/6.2	4.0/10.2
Maximum cladding wall temperature (°F)	1,170	1,215	1,200
Linear power rating, peak/ave (kW/ft)	14/7.6	14.5/7	16/11
Peak fuel burnup (MWd/Te)	80,000	150,000	150,000
Fuel volume fraction	0.329	0.325	0.342
Breeding ratio	NA	1.2	1.25
Doubling time (yrs)	NA	23	12/15
Refueling type	Offset arm	Straight pull	Offset arm
Vessel diameter/length (ft)	20.7/43.8	20.7/53.8	23.5/59.5
Primary Heat Transport System			
Reactor outlet temperature (°F)	1,050	995	1,000
Reactor inlet temperature (°F)	685	730	725
System flow rate (total 10^6 lb/hr)	18.3	41.5	136.8
Pump flow rate at pump temperature (gpm)	14,500	33,700	122,400
Pump developed heat at design flow (ft. Na)	500	450	500
Pump location	Hot leg	Hot leg	Cold leg
Intermediate Heat Transport System			
Hot leg temperature (°F)	950	936	935
Cold leg temperature (°F)	600	651	650
System flow rate (Total 10^6 lb/hr)	19.3	38.3	131.4
Pump flow rate at pump temperature	14,500	29,500	101,100
Pump developed heat at design flow (ft. Na)	400	410	475
Pump location	Cold leg	Cold leg	Cold leg

Source: Adapted from "Report of the Liquid Metal Fast Breeder Reactor Program Review Group," U.S. ERDA report, ERDA-1 (January 1975).

and a doubling time of somewhat more than 20 years. In designing the CRBR, it has been found that it is difficult to achieve a ratio of 1.2 with the conventional division between core and blanket, but that — with a heterogeneous design, mixing fissile and fertile assemblies — the original goal is feasible.

Most LMFBR development has presumed a uranium-plutonium fuel cycle.

Because the thorium-uranium cycle is superior in thermal reactors and satisfactory in fast ones, there may be advantages in introducing thorium, and possibly ^{233}U, into fast reactors, particularly where a mixed system of fast reactors and high-conversion thermal reactors is anticipated. As discussed in Chapter 12, circulation of ^{233}U, replacing most of the ^{239}Pu, may reduce the ease of diversion to weapons use.

While maintaining the basic division of the LMFBR fuel region into a core and a blanket, there are a number of possible ways to introduce thorium or ^{233}U. Thorium may be used in the blanket or the core, and ^{233}U may replace part or all of the core ^{239}Pu. Moreover, properties of thorium in metallic form may permit metallic, rather than ceramic, fuel rods. Absence of the oxygen (or carbon) would improve the nuclear properties, even raising the breeding ratio above that possible in a mixed oxide fueled LMFBR. In addition, the presence of thorium lowers the void coefficient and the Doppler coefficient, so that development of voids in the coolant and rises in temperature tend to reduce the neutron multiplication factor, not raise it.

Table 13-2 presents results of performance calculations for uranium, plutonium, and thorium-fueled LMFBRs. Results for both the initial core and an equilibrium core are shown. Several of the combinations give breeding ratios higher than a conventional LMFBR (Case 9), particularly those with metal fuel. Of particular interest if the nuclear system uses satellite thermal reactors are Cases 7 and 8, which produce substantial quantities of ^{233}U on receipt of relatively small amounts of plutonium. It is important to emphasize, though, that these possibilities are ideas that are far from the demonstration stage.

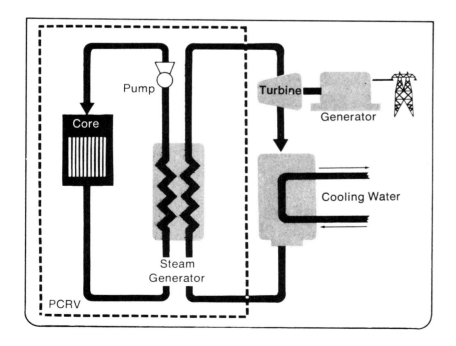

Figure 13-6. SCHEMATIC GAS-COOLED FAST REACTOR SYSTEM.
The layout for a gas-cooled fast breeder reactor is very similar to that of a high-temperature gas-cooled reactor, except that the core of the fast reactor is similar to that of an LMFBR. The primary coolant is pressurized gas, and the core, steam generators, and circulators are contained in a prestressed concrete reactor vessel (PCRV). (Figure reproduced from ERDA-76-107.)

TABLE 13-2.
Liquid Metal Fast Breeder Reactor Performance for Alternative Fuels (total power = 2500 MWth)

Case No.	Core Fuel Material	Radial Blanket Material	Core Volume (liters)	Cycle Length (days)	Total[a] Core Fissile Mass (kg)	Ratio of Fissile[b] Production to Fissile Absorption	Extra Fissile Material Production (kg/cycle)				%ΔK for Na[c] Voiding (whole core)	Doppler[c] Coefficient (whole core)
							233U	235U	239Pu	241Pu		
Equilibrium core												
1	ThO_2–$^{235}UO_2$	ThO_2	6,795	223	3,920.5	0.84	477.1	−588.5	–	–	–	–
2	ThO_2–$^{233}UO_2$	ThO_2	6,795	225	2,569.2	1.05	38.7	–	–	–	–	–
3	ThO_2–$^{239}PuO_2$	UO_2	6,795	228	2,958.7	1.15	434.3	–	−341.6	3.8	–	–
4	Th–^{235}U	Th	5,630	262	4,010.0	0.96	584.0	−619.4	–	–	–	–
5	Th–^{233}U	Th	5,630	261	2,645.3	1.14	108.0	–	–	–	–	–
6	Th–^{233}U	U	5,630	265	2,584.7	1.21	−44.1	–	210.1	–	–	–
7	Th–^{233}U–^{239}Pu	U	5,630	280	2,695.7	1.31	335.8	–	−104.9	1.7	–	–
8	Th–^{233}U–Pu[d]	U	5,630	271	2,502.6	1.34	266.8	–	31.3	−48.9	–	–
9	UO_2–$^{239}PuO_2$	UO_2	6,795	256	2,950.8	1.23	–	−31.3	165.1	7.6	–	–
Initial core												
1	ThO_2–$^{235}UO_2$	ThO_2	6,795	223	4,202.0	0.83	563.5	−685.8	–	–	−1.89	−0.0078[e]
2	ThO_2–$^{233}UO_2$	ThO_2	6,795	226	2,611.7	1.08	59.8	–	–	–	−1.98	−0.0107
3	ThO_2–$^{239}PuO_2$	UO_2	6,795	228	3,084.7	1.21	587.1	–	−460.7	2.1	0.62	−0.0093
4	Th–^{235}U	Th	5,630	262	4,384.4	0.94	707.5	−755.3	–	–	−2.52	−0.0068[e]
5	Th–^{233}U	Th	5,630	261	2,738.5	1.17	134.4	–	–	–	−2.58	−0.0086
6	Th–^{233}U	U	5,630	265	2,674.3	1.22	−37.7	–	216.3	–	−2.25	−0.0087
7	Th–^{233}U–^{239}Pu	U	5,630	280	2,817.4	1.36	458.6	–	−191.6	1.0	−0.52	−0.0084
8	Th–^{233}U–Pu[d]	U	5,630	271	2,610.6	1.39	358.8	–	−11.54	−66.4	−0.73	−0.0089
9	UO_2–$^{239}PuO_2$	UO_2	6,795	256	2,988.2	1.30	–	−40.7	211.5	3.3	3.25	–

a. Beginning of cycle values.
b. Middle of cycle values.
c. These values apply to the start of the first core cycle.
d. Contains LWR-generated plutonium.
e. Values are those of thorium only.

SOURCE: Seghal et al. Unfortunately, the table does not include a case where ^{239}Pu is the fissile material and Th is the fertile material for both core and blanket.

TABLE 13-3.

Comparison of Gas-Cooled Fast Breeder Reactor and Liquid Metal Fast Breeder Reactor Demonstration Plant Core Characteristics

	GCFR	LMFBR[a]
Thermal rating (MWth)	830	975
Electrical rating (MWe) net	300	350
Net plant efficiency (percent)	36.0%	36.0%
Fuel type	Mixed oxide	Mixed oxide
Primary coolant	Helium	Sodium
Secondary coolant	Steam/water	Steam/water
Intermediate coolant	None	None
Core inlet temperature, °F	613	720
Core outlet temperature, °F	1,022	980
Fuel elements		
Rod diameter (in.)	0.286	0.23
Cladding thickness (in.)	0.019	0.015
Cladding and duct material	Stainless steel	Stainless steel
Core height (in.)	39.2	36.0
Number of rods per element	270	217
Number of fuel and control elements	118	198
Average power density (kW/liter)	240	350
Peak fuel burnup (MWd/Te)	100,000	80,000
Peak linear power (kW/ft)	12.5	15.5
Maximum cladding temperature, mid-wall (°F)	1,268	1,215
Residence time, full power days	750	367 (peak) 506 (avg)
Peak total flux (n/cm^2–sec)	6×10^{15}	8×10^{15}
Fission-gas pressure control	Venting through pressure equalization system	Fuel-rod plenum spaces
Radial blanket elements		
Fertile material	ThO_2[b]	UO_2
Number	147	150
Rods per assembly	126	61
Breeding ratio	1.4	1.15

a. R. V. Vijuk, "Review of Core Fuel and Radial Blanket Designs for the LMBFR Demonstration Plant," Meeting at HEDL, October 24-25, 1973.

b. UO_2 is an alternative material.

Source: Gas-Cooled Fast Breeder Reactor Preliminary Safety Information Document.

GAS-COOLED FAST BREEDER REACTORS

The gas-cooled fast reactor (GCFR) is neutronically similar to the LMFBR, but externally similar to the high-temperature gas-cooled reactor. It would use a gas coolant and a large prestressed concrete reactor vessel, as the HTGR does. The heat transfer arrangement (Figure 13-6) is similar to that of an HTGR. However, the fuel and materials technology and the basic core configuration are similar to those of an LMFBR.

The most substantial difference in the core, more a matter of choice than of necessity, is that the GCFR uses thorium in the blanket region. Otherwise, the fuel design is similar to that of an LMFBR. A comparison between representative parameters for GCFR and LMFBR demonstration plants is made in Table 13-3. Because

GCFR CORE

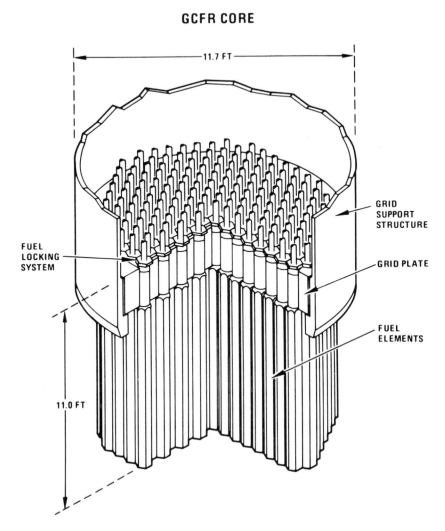

Figure 13-7. GAS-COOLED FAST REACTOR CORE.
The core of a demonstration gas-cooled fast breeder reactor (see Figure 13-8) is shown schematically. The fuel rods and assemblies are generally similar to those of an LMFBR, with the important difference that the GCFR has fuel rods whose surface is roughened. (Figure courtesy of General Atomic Co.)

of the thorium blanket, the GCFR produces an excess of ^{233}U, which could most suitably be used in a high-conversion thermal reactor, such as the HTGR.

As seen in Table 13-3, the demonstration GCFR has a substantially larger breeding ratio than the demonstration LMFBR. This is attributed, at least in part, to the fact that helium absorbs fewer neutrons, and moderates them less (because of the low-density helium), than sodium. This fact also permits a core structure that is slightly more open than that of an LMFBR, enough to eliminate the need for

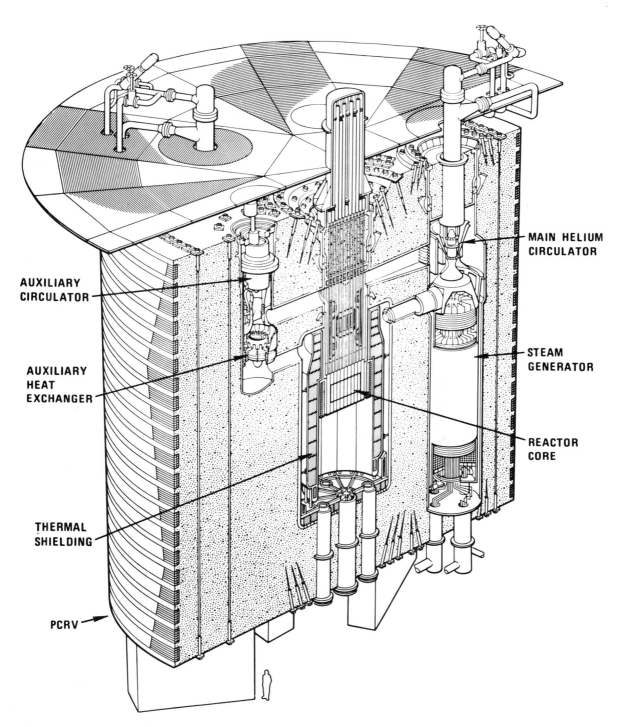

AUXILIARY
CIRCULATOR

AUXILIARY
HEAT
EXCHANGER

THERMAL
SHIELDING

PCRV

MAIN HELIUM
CIRCULATOR

STEAM
GENERATOR

REACTOR
CORE

Figure 13-8. GCFR PRESTRESSED CONCRETE REACTOR VESSEL ARRANGEMENT.
The primary system of a GCFR is contained within a large concrete vessel. Various components
are located in cavities within the vessel, and penetrations into the vessel are used for servicing
these components. (Figure courtesy of General Atomic Co.)

restraints on movement of fuel assemblies that might be caused by radiation-induced swelling of the assembly material. Nevertheless, the GCFR core is essentially as compact as that of an LMFBR. The general appearance of a GCFR core is shown in Figure 13-7.

The arrangement in the prestressed concrete reactor vessel for a demonstration GCFR is shown in Figure 13-8. Relative to an HTGR, the core occupies a relatively small portion of the PCRV. The high power density characteristic of such a core requires that special steps be taken to ensure that the helium coolant of the GCFR successfully cool the core. Unlike the LMFBR, where the sodium remains even if pressure is lost, the GCFR requires high-pressure coolant. (In contrast, the HTGR has such a large core mass that coolant has to be lost for a long period before the fuel is damaged.) Designs are incorporated at PCRV penetrations to guarantee that no large amount of the coolant ever escapes. Moreover, the basic fuel design utilizes a fuel pin with a surface that is roughened so that heat transfer between the cladding and the coolant is improved.

The fact that helium is used as a coolant obviously eliminates the difficulty of handling sodium. An intermediate cooling loop is no longer necessary. Moreover, the helium, because it is essentially transparent to neutrons, cannot cause changes in reactivity (as the sodium could, if a void developed in the coolant). On the other hand, a sodium-cooled reactor has a large amount of sodium surrounding the core; this sodium can absorb heat, relying in part even on natural convection, if pumping capability is lost. In the GCFR, forced cooling must always be provided.

The GCFR has been the object of development programs in the United States (by General Atomic) and elsewhere. No demonstration is now planned. However, the general reexamination of possible reactor types has included the GCFR in its consideration.

Bibliography — Chapter Thirteen

"NTIS" indicates report is available from: National Technical Information Service, U.S. Department of Commerce, 5285 Port Royal Road, Springfield, VA 22161.

Chang, Y. I., et al., "Alternate Fuel Cycle Options: Performance Characteristics and Impact on Nuclear Power Growth Potential," Argonne National Laboratory report ANL-77-70 (September 1977) (NTIS).

Includes calculations for LMFBRs operating on various fuel cycles, including those utilizing thorium.

CONF-740501. "Gas-Cooled Reactors: HTGR's and GCFBR's," topical conference, Gatlinburg, May 7, 1974 (NTIS).

Technical papers on gas-cooled reactors, including the gas-cooled fast breeder.

EPRI NP-142. "Development Status and Operational Features of the High Temperature Gas-Cooled Reactor," Electric Power Research Institute report EPRI NP-142 (April 1976) (NTIS).

Includes a brief summary of the status of the gas-cooled fast breeder reactor.

ERDA-1. "Report of the Liquid Metal Fast Breeder Reactor Program Review Group," U.S. ERDA report ERDA-1 (January 1975) (NTIS).

Reviews the need for the LMFBR and the progress of the program as of late 1974.

ERDA-1535. "Final Environmental Statement, Liquid Metal Fast Breeder Reactor Program," 3 vols., U.S. ERDA report ERDA-1535 (December 1975), with "Proposed Final Environmental Statement," 7 vols., U.S. AEC report WASH-1535 (December 1974) (NTIS).

Considers environmental aspects of liquid metal fast breeder reactors.

ERDA-76-107. "Advanced Nuclear Reactors," U.S. ERDA report ERDA-76-107 (May 1976) (NTIS).

A very brief introduction to the advanced reactors being considered by ERDA during 1976.

"Gas-Cooled Fast Breeder Reactor, Preliminary Safety Information Document," General Atomic Company submission to the Atomic Energy Commission (Directorate of Licensing) (as amended June 1974).

Safety description of the GCFR, comparable to the safety Analysis Report ordinarily submitted in licensing applications.

Kasten, P. R., et al., "Assessment of the Thorium Cycles in Power Reactors," Oak Ridge National Laboratory report ORNL-TM-5565 (January 1977) (NTIS).

Evaluates the potential of thorium fuel cycles.

Metz, W. D., "European Breeders (I): France Leads the Way," *Science,* vol. 190, p. 1279 (1975); "European Breeders (II): The Nuclear Parts Are Not the Problem," *Science,* vol. 191, p. 368 (1976); "European Breeders (III): Fuels and Fuel Cycles are Keys to Economy," *Science,* vol. 191, p. 551 (1976).

Describes the European efforts to build liquid metal fast breeder reactors.

Pigford, T. H., et al., "Fuel Cycles for Electric Power Generation," Teknekron report EEED 101 (January 1973, rev. March 1975), part of "Comprehensive Standards: The Power Generation Case," U.S. EPA report PB-259-876 (March 1975) (NTIS).

Estimates nuclear material flows and environmental releases for various electrical generating systems, including the LMFBR.

Pigford, T. H., and Ang, K. P., "The Plutonium Fuel Cycles," *Health Physics,* vol. 29, p. 451 (1975).

Calculates the flow of nuclear material for a number of plutonium fuel cycles, including the LMFBR.

"Preapplication Safety Evaluation of the GCFBR," U.S. AEC (licensing project No. 456; August 1, 1974).

Evaluation of the safety aspects of the GCFR, based on preliminary information supplied by General Atomic Company.

Seaborg, G. T., and Bloom, J. L., "Fast Breeder Reactors," *Scientific American,* vol. 223, p. 13 (November 1970).

Describes LMFBRs and GCFRs.

Seghal, B. R., Lin, C. L., and Naser, J., "Performance of Various Thorium Fuel Cycles in LMFBRs," Electric Power Research Institute, *EPRI Journal,* vol. 2, p. 40 (September 1977).

Calculates input and output of nuclear materials to and from LMFBRs operated on thorium cycles.

Vendryes, G. A., "Superphenix: A Full-Scale Breeder Reactor," *Scientific American,* vol. 236, p. 26 (March 1977).

Elementary description of the next generation of the French LMFBR.

WASH-1089. "An Evaluation of Gas-Cooled Fast Reactors," U.S. AEC report WASH-1089 (1969) (NTIS).

An early AEC evaluation of gas-cooled fast reactors.

WASH-1102. "LMFBR Program Plan, Element 2, Plant Design," 2nd ed., U.S. AEC report WASH-1102 (December 1973) (NTIS).

Describes design of liquid metal fast breeder reactors.

CHAPTER FOURTEEN

Advanced Thermal Reactors: The Thorium-Uranium Economy

THE URANIUM-PLUTONIUM fuel cycle, as applied in thermal reactors, has severe limitations. As has been noted throughout this book, ^{239}Pu, the product of neutron capture by ^{238}U, simply does not yield enough neutrons from thermally induced fission to break even. Some ^{239}Pu can be produced, but not enough to replace that which is destroyed. In practice, conversion ratios of 0.6 have been typical in commercial reactors. Strong attention to the conversion ratio would doubtless yield some improvement, but the neutronic limitation is severe.

Nonetheless, such attention may be worth the effort, if only to enlarge our awareness of the full range of possibilities in thermal reactor systems. As noted earlier, more detailed fuel management and even introduction of some heavy water into LWRs (as in a "spectral shift control reactor") could raise the overall conversion ratio. The spectral shift control reactor, in fact, would owe its superiority, not so much to the lower absorptive properties of heavy water, but to the fact that undermoderating the neutron spectrum raises the rate of capture by ^{238}U (or ^{232}Th), producing ^{239}Pu (or ^{233}U). With the relatively small coolant volume typical of LWRs, as compared to HWRs, the available deuterium does not moderate as effectively as would ordinary hydrogen. The result is that neutrons are not lost to control poisons.

Interestingly enough, a comparable approach is used in the light-water breeder reactor (LWBR), discussed later in this chapter, to reduce losses to neutron poisons; however, the LWBR uses a variable geometry rather than a variable moderator, to vary the balance between fission and capture.

A more fundamental difference, however, is that the LWBR operates on a thorium-uranium fuel cycle. It is such reactors that are the main focus of this chapter. Many of the important features of thorium-uranium neutronics were indicated in Chapter 10 and in Appendix C. We note below some general techniques for improving neutron utilization in thermal reactors.

220

A fundamental step in making such improvement is simply to use thorium and ^{233}U. A beginning has been made in investigating the possibility of introducing this fuel cycle into LWRs operating now and in the future; the LWBR would be a more advanced form of such reactors, optimizing neutron utilization to achieve a barely breeding condition. The heavy-water CANDU reactor, too, is adaptable to a thorium-uranium cycle; this possibility has long been in the minds of Canadian developers and probably has a greater potential for break-even than light-water reactors. The high-temperature gas-cooled reactor has already been marketed with a thorium cycle (and subsequently withdrawn); there has lately been interest in various possible fuel loadings for the HTGR, and those that are most resource efficient involve thorium cycles. Finally, an even more advanced form of thermal reactor, as far as uranium utilization is concerned, is the molten salt breeder reactor (MSBR), which is widely believed to have the capability of breeding; it is, however, not being developed.

This chapter is primarily concerned with thermal reactors with improved efficiency of uranium utilization. Typically, they require fuel reprocessing and recycle of fissile material. The possibility of improving uranium utilization in once-through fuel cycles (i.e., no recycle) is discussed briefly in Chapter 12, where means to minimize the probability of weapons proliferation are treated.

NEUTRON ECONOMIES FOR THERMAL REACTORS

The discussion of reactor types that follows should be considered in the light of the several means available for improving the conversion ratio in solid fuel reactors. (Some of the same means pertain to the molten salt reactor and other concepts.) The conversion ratio for any given reactor results from a delicate balance: available neutrons are absorbed by fissile, fertile, and other materials. The last category of absorptions may be regarded as parasitic, although a portion of them may serve a useful purpose in the form of poison control. Conversion may be improved by shifting the distribution of neutrons among these three types of material or by increasing the number of neutrons available.

To begin with, use of the thorium-uranium cycle generally increases the number of available neutrons above that from uranium-plutonium, largely because the neutron yield per absorbed thermal neutron is greater for ^{233}U than for ^{239}Pu. (To be more precise than this, the detailed neutron balance would have to be calculated for each system of interest.) As a result, introduction of thorium into a previously uranium system usually reduces demand on fissile resources, provided the comparison is made between complete fuel cycles. Reoptimization for the new fuel often reduces demand further.

There are, on the other hand, several opportunities for reduction of parasitic neutron absorptions. One is to choose a moderator that absorbs fewer neutrons, say heavy water instead of light water. However, as was apparent from Table 5-2, the largest opportunity for reducing absorptions in an LWR is to reduce poison control. This is, in fact, probably the major reason for the higher conversion ratio of a CANDU, as compared to an LWR. This improvement in the CANDU is accomplished by on-line refueling, which reduces the need for control poisons by eliminating the change of reactivity that, in an LWR, occurs as the amount of fissile material decreases during the refueling cycle.

As mentioned above, two other methods for substantially reducing the amount of poison control are spectrum control, as would be employed in a spectral shift control reactor, and a variable geometry, as used in an LWBR. Both of these measures alter the balance of neutron absorption by fissile and fertile materials to compensate for reactivity changes during the refueling cycle.

In all the concepts discussed below, thorium is the principal fertile material, although some ^{238}U is often involved. In addition, one or more of the measures for reducing parasitic neutron absorptions is typically adopted.

One incidental result of thorium utilization is that, because thorium tends to capture more neutrons than ^{238}U, maintenance of a critical system requires a higher fissile loading than in a uranium fueled reactor. And as the conversion ratio is increased by alteration of fuel design or management, the fissile commitment necessary to reach an equilibrium state increases. Consequently, the initial fuel costs tend to concentrate more and more at the reactor's startup until, for break-even (conversion ratio 1), fissile commitments are all at the beginning of life. This is a distinct disadvantage as far as current economics is concerned but may actually be an advantage as far as resource utilization is concerned, forcing the uranium industry into a mode that makes for most effective extraction of available resources.

THORIUM FUEL IN LIGHT-WATER REACTORS

Although current light-water reactors use a low-enrichment uranium-plutonium fuel cycle, it is clearly possible to employ the thorium-uranium cycle in a similar environment. Considering how widespread LWR technology is, this could provide a way of rapidly progressing to a more advanced technology, primarily with the purpose of improving uranium utilization. Either new reactors, based on the LWR industry, could be designed, or old reactors could be refitted.

The latter possibility is, naturally, the easiest, particularly if the only substantial change is to alter the fuel composition. Preliminary calculations indicate that the best PWR thorium-uranium performance is obtained if the reactor is fueled with thorium and highly enriched uranium including recycled ^{233}U. The equilibrium demand on resources by a 1000-MWe plant (75% capacity factor) is then about 90 tons of U_3O_8 per year. (This is similar to the value for a standard HTGR.) In addition, a substantial commitment must be made at the beginning of life, yielding a lifetime commitment of about 3500 tons, noticeably better than the 4000 tons associated with plutonium recycle. This assumes a fuel burnup and cycle time similar to that for a conventional PWR. The conversion ratio would be around 0.75, noticeably higher than in ordinary LWRs.

Fuel and operational design modifications may lead to further improvement. One such modification is to reduce the coolant-to-fuel ratio, thereby decreasing neutron absorption by water. This appears to lead to a decrease in the annual uranium feed requirement by about 15%, although, as usual, this decrease is accompanied by an increase in initial fissile commitment. A lower burnup fuel cycle also decreases annual makeup demands slightly, by reducing the reactivity swing and its dampening by control poisons.

As in the uranium-plutonium reactors, losses to control poisons may be largely eliminated by reliance on spectral shift control. This concept has not been investigated in detail, but it appears that conversion ratios between 0.8 and 0.9 may be achieved, implying annual feed requirements of about 50 tons of U_3O_8. This approaches break-even. The next step is the LWBR, which would achieve conversion ratios of about 1.0.

THE LIGHT-WATER BREEDER REACTOR

The light-water breeder reactor is now being developed by the Department of Energy (formerly ERDA) on a rather low-level basis. It consists basically of a pressurized water reactor in which the core runs on a thorium-uranium cycle, much as discussed above. The novel feature of the LWBR, aside from its dependence on the thorium cycle, is the lack of neutron poisons as control. Although boron poisons are held in reserve as a backup shutdown system, they are not used under normal operating conditions. The reactor is controlled by moving fuel assemblies with higher than average enrichment partially in and out of the core. This changes the core configuration, and hence the neutron multiplication factor, and avoids the neutron loss caused by control poisons. Withdrawal of the high-enrichment (i.e., about 6%) fuel assemblies controls the ratio of fissions to conversions, and can be used to maintain a multiplication factor of 1, as required for constant-power operation. Overall, the LWBR core is a much more highly optimized system than that discussed above in the context of modified PWRs.

Figure 14-1 shows a test PWR reactor vessel as modified for testing an LWBR core. (The reactor vessel is part of the Shippingport plant, designed for producing much less power than the full-scale 1000-MWe plants of which we have been speaking.) Since an LWBR is simply a modification of a PWR system, one should envision such a reactor vessel replacing the vessel in an ordinary PWR system (Figure 5-1). The modifications required on the vessel are primarily a new core, vessel head, and control system. (Adapting a BWR to this concept would be different because of the different water conditions and because the control system consists of plates manipulated from the bottom, see Chapter 6.) In Figure 14-1, note the large mechanisms for moving the "seed" (i.e., higher enrichment) regions; these mechanisms raise seed regions into the core to increase the multiplication factor.

The fuel assemblies themselves are not wholly dissimilar from those of an LWR, consisting of dioxides formed into cylindrical ceramic pellets sealed into cladding which, in this case, is a zirconium alloy which absorbs particularly few neutrons. Fuel rods will contain different mixtures of oxides, depending on the core region shown in Figure 14-2. The "seed" and "blanket" regions initially contain a mixture of $^{233}UO_2$ and ThO_2, the seed region having up to 6% uranium and the blanket regions about 3%. The reflector region (analogous to the so-called "blanket" region of a fast breeder) intially contains all thorium, part of which is

Figure 14-1. LIGHT-WATER
REACTOR BREEDER CORE
AND VESSEL.
A light-water reactor system is a
pressurized-water reactor in which
the core is modified to achieve a conversion ratio of approximately 1.
This is accomplished by using thorium ^{233}U fuel with a low moderator-
to-fuel ratio and by controlling the
reaction rate, not with control rods,
but by moving parts of the core. The
configuration is changed by moving
entire fuel assemblies. (The figure
shows the Shippingport test reactor
as modified for LWBR testing and is
reproduced from ERDA-1541.)

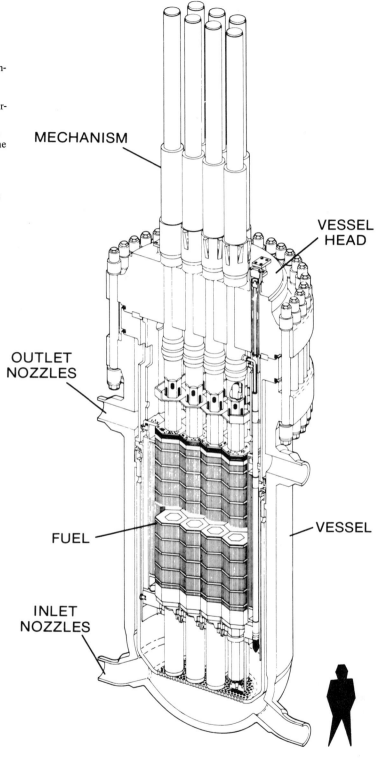

MECHANISM

VESSEL
HEAD

OUTLET
NOZZLES

FUEL

VESSEL

INLET
NOZZLES

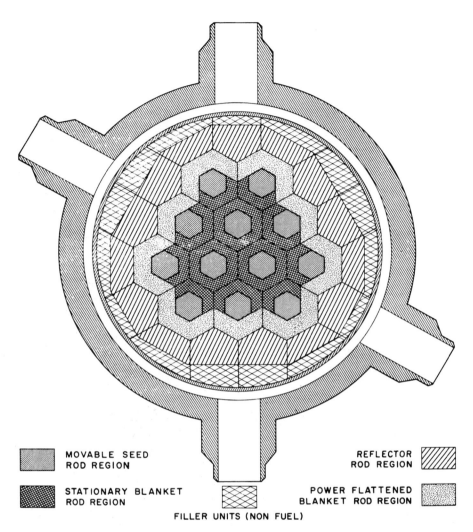

	MOVABLE SEED ROD REGION		REFLECTOR ROD REGION
	STATIONARY BLANKET ROD REGION		POWER FLATTENED BLANKET ROD REGION

FILLER UNITS (NON FUEL)

Figure 14-2. LIGHT-WATER BREEDER REACTOR TEST CORE CROSS-SECTION. An LWBR core consists basically of moderately enriched movable "seed" regions, surrounded by blanket regions of low enrichment. These are both surrounded by fertile material. (The figure shows the core regions for the Shippingport test reactor and is reproduced from ERDA-1541.)

converted to uranium as the reactor operates. It is claimed that this configuration will breed, although quite slowly, with a ratio exceeding 1 by no more than a few percent.

Much as in the case of thorium CANDU, getting to the high conversion or breeding situation (where the core contains primarily ^{233}U and ^{232}Th) requires a "prebreeder" phase in which the ^{233}U is produced. In the case of a commercial LWBR, this phase could use a combination of thorium plus uranium highly enriched in ^{235}U to breed the ^{233}U. Achieving the breeder stage would require a total investment of roughly 2000 tons of U_3O_8 per 1000 MWe reactor, after which, as noted in Chapter 10, the bred ^{233}U could fuel an LWBR essentially forever.

The Shippingport plant is being used as a test of LWBR feasibility. During 1977, it began operating with an LWBR core, and results should be available around 1980. This plant is, however, not a demonstration in the usual sense (as the Clinch River breeder would be). Such a demonstration would follow the Shippingport experiment.

ADVANCED HEAVY-WATER REACTORS

There are a variety of ways to incorporate heavy water moderation into a reactor system. The Atomic Energy of Canada, Ltd. markets a single type, a pressurized heavy water cooled reactor with pressure tube construction. England's nuclear development agencies have designed a "steam generating heavy-water reactor," which has a moderator vessel (calandria) like the CANDU, but which has vertical fuel channels filled with light water, which is permitted to boil. The SGHWR is thus a boiling water (cooled) heavy water (moderated) reactor. This reactor has not reached commercial status. As a matter of fact, the AECL regards boiling water cooling to be a possibility for the CANDU. A third possibility for coolant is an organic fluid, which would have the advantage of operating at higher temperatures, thus yielding greater thermal efficiencies. It does not, however, appear likely that AECL will develop either of these concepts.

A more likely prospect is to begin building thorium-fueled CANDUs later this century. This offers the possibility of a substantial decrease in uranium requirements. In principle, a thorium system could be started either with the plutonium present in irradiated uranium fuel or with the ^{235}U present in natural uranium resources. The latter approach turns out to be most resource efficient.

The plutonium now residing in spent CANDU fuel turns out to be a very expensive "resource," largely because reprocessing the fuel would yield only the 0.2% to 0.3% of fissile plutonium. The uranium is so depleted in ^{235}U (about 0.2%) that it is comparable to enrichment tails. If CANDU fuel were to be reprocessed, and the plutonium recycled to the reactors that produced it, the resource requirements would be decreased by almost half, making the CANDU the most resource efficient thermal system in existence. On the other hand, feeding it to thorium reactors would build up the required ^{233}U inventory, a valuable use. However, reprocessing the large amount of spent fuel that would be needed would be expensive.

A more direct way to build up the required ^{233}U inventory would be to feed highly enriched ^{235}U to the thorium CANDU. This obviously requires enrichment facilities. Even so, this may be less expensive than extracting plutonium, and it has the advantage that the net demand on uranium resources is decreased. At the extreme, a break-even reactor would require a commitment of about 1500 tons of U_3O_8 (the numbers cited vary from 1300 to 1700) to build up the necessary ^{233}U inventory for one 1000-MWe reactor and its fuel cycle. A system with a poorer conversion ratio would have a smaller initial requirement, but would require some annual makeup.

TABLE 14-1.

Approximate Neutronics for a High-Burnup Thorium CANDU

Approximately 2.25 fast neutrons are produced following the absorption of 1 neutron by fissile material and have the following fate:

0.87[a]	Captured by fertile material, leading to fissile production
1	Absorbed by fissile material
0.02	Absorbed by heavy water
0.24	Absorbed by fission products and structure
0.09	Absorbed by other materials, including control poisons
0.03	Lost by leakage
2.25	

---

a. The conversion ratio is therefore 0.87.

It appears that use of thorium with ^{233}U recycle immediately raises the conversion ratio to about 0.8, that moderate attention to reactor design raises it to 0.9, and that a willingness to accept low burnups (and thus recycle the fuel often) can raise it to 1.0, break-even. Table 14-1 shows the distribution of neutron absorptions for a high-burnup system. The use of thorium and ^{233}U results in a slight increase in the neutron production ratio and in the conversion ratio. Reducing the burnup to about 10,000 MWd/Te, comparable to what is used in CANDUs now, can raise the conversion ratio to 1.0. However, the mass of fuel that would require reprocessing yearly is several times the amount for an LWR. (See Table 11-1.) Economic optimization would now dictate high-burnup fuel. Resource optimization, on the other hand, would suggest paying the economic penalty to achieve high conversion.

One might ask how efficient the thorium-CANDU would be if the ^{235}U were supplied at lower enrichment, say 10% or 20%. Since most of the fuel would still be thorium, it can be expected that the conversion ratio would remain high, although slightly less than without the ^{238}U. However, the AECL has ordinarily not considered such a possibility.

At this point, it is useful to note how annual feed requirements and startup commitments compare for various systems previously discussed. Figure 14-3 gives this information for the break-even CANDU (with both plutonium and ^{235}U startup), the LWBR, and LMFBRs (with both oxide and advanced carbide fuels). The data on PWRs pertain not only to three conventional alternatives but to thorium-fueled PWRs, discussed earlier in this chapter. The smooth curve drawn through the points is empirical (with no significance of itself). Much of this information is contained in Table 10-2.

HIGH-CONVERSION GAS-COOLED REACTORS

The gas-cooled reactor described in Chapter 8 had an economic conversion ratio, less than 0.7, for the time when it was marketed. As in many reactor systems, higher ratios and lower uranium feed requirements can be achieved. As an indica-

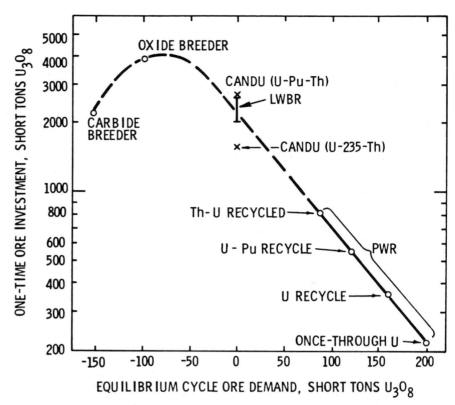

Figure 14-3. URANIUM DEMAND (STARTUP VERSUS ANNUAL MAKEUP) FOR 1 GWe
WATER-COOLED AND LIQUID-METAL-COOLED NUCLEAR PLANTS.
A nuclear plant requires an initial fissile investment for reaching equilibrium and an annual
makeup supply afterwards. The figure plots uranium requirements for initial versus makeup
fueling for several PWR types (including one fueled with thorium-uranium fuel), for break-even
CANDUs, and for LMFBRs. (Figure reproduced from EPRI NP-359.)

tion of what can be done, Table 14-2 shows the neutron balance for a relatively
high conversion reactor of the HTGR type. The conversion ratio achieved is slightly
above 0.8. Even higher ratios may be possible but, as usual, at the expense of more
frequent refueling and more fuel reprocessing.

The HTGR has generally been thought of as a ^{235}U–thorium–^{233}U system,
where the initial feed is 93% ^{235}U. Some analyses of the effect of using lower
enrichments, chosen to preclude weapons use, have been performed. As expected,
there is some degradation of performance, but, as usual, it is not startling. It has
been suggested, in any case, that the great mass of the HTGR core would serve as a
deterrent to theft by terrorists. (A lot of material would have to be transported to
carry off a critical mass.)

The HTGR is in the odd position of no longer being available commercially,
but of drawing considerable attention nonetheless. General Atomic, the United
States manufacturer, is persistent in its attempts to arouse interest in its advantages.
Other countries also maintain a moderate interest in gas-cooled reactors. One such

concept is the "pebble bed" reactor, similar to the HTGR, except that the core is not stacked blocks of carbon with fuel loadings. Instead, the core consists of small spherical fuel/carbon balls, a few centimeters in diameter. The reactor vessel contains these fuel balls. The reactor can be refueled on-line, simply by removal of some of the balls at the bottom as new ones are added on top. The pebble bed reactor is being developed in Germany. In addition, England has constructed gas-cooled reactors with a carbon dioxide coolant.

THE MOLTEN SALT BREEDER REACTOR

Relative to the other reactors we have discussed, the molten salt breeder reactor is the least developed. Test reactors have run and, until 1976, there had been a place for the MSBR in the AEC or ERDA budget. However, at the present time, no government agency or commercial vendor is seriously developing the concept. We discuss it here because the concept is often mentioned in the context of advanced systems, and it may offer the highest breeding ratio in a thermal system.

The concept of a fluid reacting material is radically different from the other systems that we have discussed. In such a reactor, the "core" is simply that point in the primary loop where a critical mass assembles. This point occurs, not surprisingly, in the reactor vessel. Fluid is withdrawn from this vessel for purposes of extracting heat for generating power and for reprocessing purposes, i.e., to remove from the system neutron poisons that would reduce the breeding ratio. A basic conception of the MSBR is shown in Figure 14-4. The reactor fluid is fissile uranium and fertile thorium as tetrafluorides, dissolved in a carrier salt of lithium and beryllium fluoride. At the indicated heat exchangers, heat is transferred from the primary fluid to a sodium fluoroborate fluid which runs through the steam generators. Molten salt enters at the bottom of the reactor vessel and is pumped upward in a once-through arrangement. The internals of the vessel are defined by graphite moderator elements. These direct the flow patterns and also determine the fuel-moderator composition throughout the vessel. The ratio of salt to carbon can vary

TABLE 14-2.
Approximate Neutronics for a High-Conversion Gas-Cooled Carbon-Moderated Reactor

Approximately 2.2 fast neutrons are produced following the absorption of 1 neutron by fissile material and have the following fate:

0.83[a]	Captured by fertile material, leading to fissile production
1	Absorbed by fissile material
0.04	Absorbed by carbon
0.18	Absorbed by fission products
0.08	Absorbed by control poisons and lost by leakage
0.05	Absorbed by other
2.18	

a. The conversion ratio is therefore 0.83.

from one part to another. In one concept, the central portion was 19% salt (as opposed to graphite), the next 17%, and the outer 44%; the salt was, measured in moles, 68% LiF, 20% BeF_2, 12% ThF_4, and 1/3% $^{233}UF_4$. The resulting breeding ratio was estimated to be 1.05. Some important parameters for one version of an MSBR are given in Table 14-3.

A basic requirement of this reactor is that methods for extracting various chemicals to prevent deterioration of the breeding ratio be found. It is particularly important to prevent buildup of protactinium 233, the intermediate step in the decay of ^{233}Th (formed by ^{232}Th + neutron) to ^{233}U. If ^{233}Pa, which has a 27-day half-life, captures a neutron, not only is a neutron wasted, but a ^{233}U nucleus that would have been formed is not. The possibility of such a double loss exists for any thorium-uranium system, but if it is intended that the system actually

Figure 14-4. SCHEMATIC MOLTEN SALT BREEDER REACTOR.
In a molten salt breeder reactor, a fuel salt, bearing the nuclear fuel materials, circulates in a primary loop that includes the core and heat exchangers. A secondary salt loop carries heat from this exchanger to steam generators, where steam is formed for driving the turbogenerators. (Figure courtesy of Oak Ridge National Laboratory.)

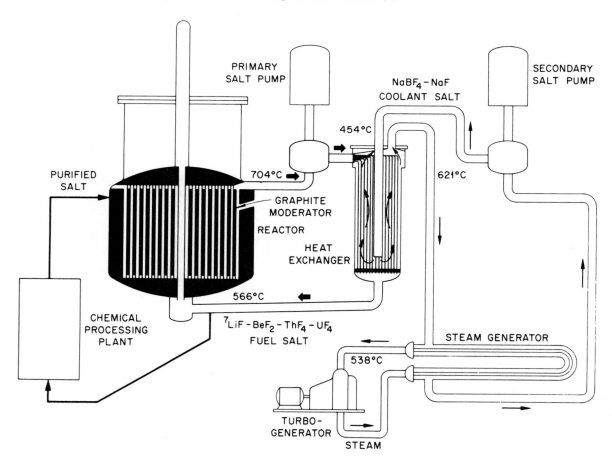

TABLE 14-3.
Principal Operating Parameters of a 1000-MWe Molten
Salt Breeder Reactor

General

Thermal power	2,250 MWth
Electric power	1,000 MWe
Thermal efficiency	44%
Fuel processing scheme	On-line, continuous processing
Breeding ratio	1.07
Fissile inventory	1500 kg
Doubling time	19 years

Reactor

Fuel salt	^{7}LiF-BeF$_2$-ThF$_4$-UF$_4$
Moderator	Unclad, sealed graphite
Reactor vessel material	Modified Hastelloy-N
Power density	22 kW/liter
Exit temperature	1300 °F (704 °C)
Temperature rise across core	250 °F (139 °C)
Reactor vessel height	20 ft (6.1 m)
Reactor vessel diameter	22 ft (6.7 m)
Vessel design pressure	75 psi (0.5 MPa)

Source: Abstracted from ERDA-1535.

breed, such losses must be given very careful attention. In the MSBR this double loss can be prevented, for the most part, by rapid processing directly from the salt. The proposed solution is a relatively delicate liquid-bismuth extraction process. Basically, liquid bismuth comes into contact with the reactor salt, and relatively different chemical potentials should cause a favorable balance in the passage of uranium, protactinium, and fission products into the bismuth fluid (a delicate matter indeed).

This on-line reprocessing is extremely important for the MSBR. In systems discussed above (such as the CANDU), on-line refueling avoids the reactivity swing associated with batch refueling. In the MSBR, there would be a much smaller change, primarily due to buildup of fission product poisons. On-line reprocessing reduces this loss, but it is thought that prevention of neutron absorptions by ^{233}Pa is even more important. However, this requires that each reactor have its own chemical reprocessing facility.

For the design indicated in Figure 14-4 and Table 14-3, the fissile inventory, for both the reactor and reprocessing plant, was estimated to be about 1500 kg. This inventory is considerably less than that of an LMFBR (about 3000 to 4000 kg). As a result, even though the MSBR of Table 14-3 has a breeding ratio of only 1.07 (considerably less than that anticipated for a commercial LMFBR), the doubling time would be about 19 years, comparable to that of an LMFBR.

Bibliography — Chapter Fourteen

"NTIS" indicates report is available from: National Technical Information Service, U.S. Department of Commerce, 5285 Port Royal Road, Springfield, VA 22161.

Banerjee, S., Critoph, E., and Hart, R. G., "Thorium as a Nuclear Fuel for CANDU Reactors," *Canadian Journal of Chemical Engineering,* vol. 53, p. 291 (1975).
 Discusses conversion ratios and fuel utilization in thorium-fueled CANDUs.

Chang, Y. I., et al., "Alternate Fuel Cycle Options: Performance Characteristics and Impact on Nuclear Power Growth Potential," Argonne National Laboratory report ANL-77-70 (September 1977) (NTIS).
 Gives fuel utilization for various thermal reactor systems.

CONF-740501. "Gas-Cooled Reactors: HTGR's and GCFBR's," topical conference, Gatlinburg, May 7, 1974 (NTIS).
 Collection of papers on gas-cooled reactors, including HTGRs.

Dahlberg, R. C., "Benefits of the HTGR Fuel Cycle: Compilation and Summary," General Atomic Company report GA-A14398 (March 1977); M. H. Merrill, "Low Enrichment Uranium/Thorium (Denatured) Fuel Cycles and Safeguards Considerations of Alternate Cycles," addendum to GA-A14398 (June 1977).
 Suggests favorable aspects of HTGR fuel cycles, including those with low enrichment.

EPRI NP-142. "Development Status and Operational Features of the High Temperature Gas-Cooled Reactor," Electric Power Research Institute report EPRI NP-142 (April 1976) (NTIS).
 Summarizes status of the HTGR, including its application in direct cycles with higher thermal efficiency.

EPRI NP-359. "Assessment of Thorium Fuel Cycles in Pressurized Water Reactors," Electric Power Research Institute report EPRI NP-359 (February 1977) (NTIS).
 Report of calculation of fuel requirements for PWRs operating on thorium fuel cycles.

EPRI NP-365. "Study of the Developmental Status and Operational Features of Heavy Water Reactors," Electric Power Research Institute report EPRI NP-365 (February 1977) (NTIS).
 Examines the current status of CANDU reactor systems, and how they might be altered to increase conversion ratios.

ERDA-1. "Report of the Liquid Metal Fast Breeder Reactor Program Review Group," U.S. ERDA report ERDA-1 (January 1975) (NTIS).
 Evaluates the need for the LMFBR, considering the potential for other reactor types to achieve lower uranium requirements.

ERDA-1535. "Final Environmental Statement, Liquid Metal Fast Breeder Reactor Program," 3 vols., U.S. ERDA report ERDA-1535 (December 1975), with "Proposed Final Environmental Statement," 7 vols., U.S. AEC report WASH-1535 (December 1974) (NTIS).
 Includes description of alternatives to the LMFBR, such as the molten salt reactor.

ERDA-1541. "Final Environmental Statement, Light Water Breeder Reactor Program, Commercial Application of LWBR Technology," 5 vols., U.S. ERDA report ERDA-1541 (June 1976) (NTIS).
 Statement of environmental impacts, including uranium requirements, of light-water breeder reactors.

ERDA-76-107. "Advanced Nuclear Reactors," U.S. ERDA report ERDA-76-107 (May 1976) (NTIS).

A brief description of the advanced reactors being considered by ERDA during 1976.

Foster, J. S., and Critoph, E., "The Status of the Canadian Nuclear Power Program and Possible Future Strategies," *Annals of Nuclear Energy,* vol. 2, p. 689 (1975).

Gives estimated fuel requirements for and characteristics of advanced CANDU reactors.

Kasten, P. R., et al., "Assessment of the Thorium Fuel Cycle in Power Reactors," Oak Ridge National Laboratory report ORNL-TM-5565 (January 1977) (NTIS).

Considers the use of thorium-uranium fuel cycles in commercial reactors.

Merrill, M. H., "Use of the Low Enriched Uranium Cycle in the HTGR," General Atomic Company report GA-A14340 (March 1977).

Summarizes results from fuel utilization calculations for HTGRs using low-enrichment uranium fuels in a once-through mode.

Perry, A. M., and Weinberg, A. M., "Thermal Breeder Reactors," *Annual Review of Nuclear Science,* vol. 22, p. 317 (1972).

Examines the potential for thermal reactors to achieve high conversion ratios on thorium-uranium cycles.

Pigford, T. H., et al., "Fuel Cycles for Electric Power Generation," Teknekron report EEED 101 (January 1973, rev. March 1975); "Fuel Cycle for 1000-MW Uranium-Plutonium Fueled Water Reactor," Teknekron report EEED 104 (March 1975); "Fuel Cycle for 1000-MW High Temperature Gas-Cooled Reactor," Teknekron report EEED 105 (March 1975). These are part of "Comprehensive Standards: The Power Generation Case," U.S. EPA report PB-259-876 (March 1975) (NTIS).

Basic information on fuel cycles, including that of the standard HTGR.

Till, C. E., et al., "A Survey of Considerations Involved in Introducing CANDU Reactors into the United States," Argonne National Laboratory report ANL-76-132 (January 1977) (NTIS).

Considers economic and licensing aspects of CANDUs, including those on a thorium-uranium cycle.

WASH-1085. "An Evaluation of High-Temperature Gas-Cooled Reactors," U.S. AEC report WASH-1085 (December 1969) (NTIS).

AEC evaluation of HTGRs.

WASH-1097. "The Use of Thorium in Nuclear Power Reactors," U.S. AEC report WASH-1097 (June 1969) (NTIS).

A relatively early discussion of the use of thorium-uranium fuel cycles.

WASH-1222. "An Evaluation of the Molten Salt Breeder Reactor," U.S. AEC report WASH-1222 (1972) (NTIS).

Description and evaluation of the molten salt breeder reactor (see also ERDA-1535).

CHAPTER FIFTEEN

Mixed Fission and Miscellany

L EST THE IMPRESSION exist that the many reactor systems treated in earlier chapters exhaust the range of possibilities, this chapter will briefly introduce some of the more exotic and new concepts that use nuclear reactions to produce energy. Up to this point, we have treated only fission-based systems, but even within this category, the number of possibilities is tremendous. Pure fission systems include not only numerous solid-fuel reactors, a number of which are the main subject of this book, but also liquid-core reactors, such as the molten salt breeder reactor discussed previously, and gas-core reactors, which we have not treated at all. In this chapter, we broaden our discussion beyond pure fission systems to those that combine fission in one way or another with nonfission concepts, in particular with fusion reactors or with high-energy particle accelerators. The systems that combine fission with other techniques are more technically demanding, even conjectural, than the advanced pure fission concepts. On the other hand, they may have much to offer in certain circumstances. For this reason, it is appropriate to treat them at the close of our discussion.

Two general concepts will be presented in this chapter. The first will be the possibility of combining fusion devices with fission devices in a way that has some specific net benefit. The second will be the use of accelerators to breed fuel for fission systems. In some respects, these possibilities are similar. Both use an auxiliary system, in one case a fusion reactor and in the other an accelerator, to breed fissile material. However, the first uses what is clearly an exotic device, a fusion reactor. We begin by noting the basic features of possible fusion systems.

FUSION

There are a variety of ways to liberate energy during nuclear reactions, but only two basic approaches offer much hope of large-scale net energy production. The first is to split heavy nuclei into medium-weight nuclei, releasing energy. The second is to join light nuclei to form somewhat heavier ones, again releasing energy.[1] Whereas the materials of fission reactors are isotopes of uranium and plutonium, the materials of fusion devices are likely to be deuterium (D or ^{2}H) and tritium (T or ^{3}H), the heavier isotopes of hydrogen. These can release energy in the following reactions:

$$D + T \longrightarrow \; ^4He + n + 17.6 \; MeV$$

$$D + D \longrightarrow \begin{cases} T + p + 4.0 \; MeV \\ \\ ^3He + n + 3.3 \; MeV \end{cases}$$

It is possible to split nuclei at the relatively low temperatures of a (fission) nuclear reactor because fission is induced by and releases neutrons, chargeless particles that can penetrate nuclei with relative ease. However, fusion reactions require that two light nuclei, each with an electrical charge, interact in spite of the fact that their charges tend to drive them apart. Achieving a reasonable probability of interaction requires that the particles have a large kinetic energy. This condition can easily be achieved particle by particle in an accelerator (see later section). However, to gain energy via fusion, the large energy expenditure of an accelerator must be avoided, and it is preferable that a large reaction mass reach the necessary energy at a given time.

These requirements may be met in a fusion device by raising a reaction mass to very high temperatures (corresponding to the keV energies likely to be necessary for a reasonable reaction rate). Energies in the keV range may seem small, but to raise a macroscopic amount to this energy means raising its temperature to millions of degrees centigrade. In rough terms, the D-T reaction in the above equation would be achieved with earlier technologies because it requires only keV energies. The D-D reaction would require temperatures an order of magnitude higher.[2]

Two classes of devices are under development, each intent on adequately confining a reaction mass while at the same time raising its temperature. At the temperatures of interest the mass is completely ionized, forming a "plasma," so that magnetic fields can be used to contain the ions. This is the basic idea of "magnetic confinement" devices, which are being studied in several configurations. These devices use large magnets to control the orbits of particles in the plasma. An alternative approach is so-called "inertial confinement." In devices based on this idea, small masses of material are quickly raised to high temperature (actually in an inward directed explosion) and both mass and the design of the implosion tend to keep the reacting particles together, albeit for a very short time. The implosion may

---

[1] The reason that both these techniques can yield energy is that both the very light and very heavy nuclei are relatively heavy, considering the number of protons and neutrons they contain. Nuclei in the middle, on the other hand, are relatively light (actually because the protons and neutrons are more tightly bound together), so that fusion of heavy nuclei or fusion of light ones results in a net loss of mass. Since mass and energy are equivalent, conservation laws (see beginning of Appendix B) require that energy be released.

[2] Although an objective of fusion devices is to raise the temperature, the ultimate purpose is to produce energy on a practical scale. This means achieving a useful combination of temperature and density over a sufficient period of time. (The details are beyond the scope of this discussion.)

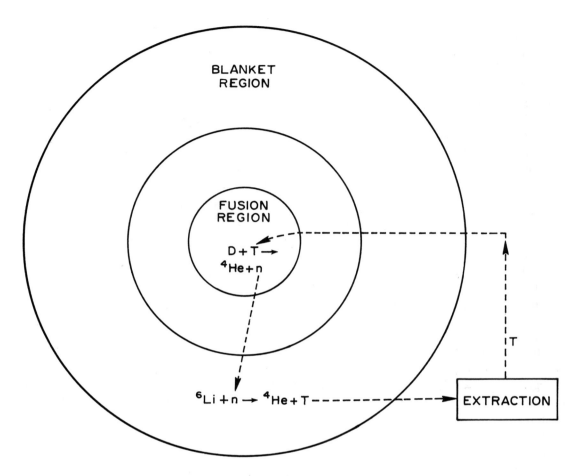

Figure 15-1. SCHEMATIC FUSION REACTOR.
In the fusion region, deuterium and tritium react, producing energetic neutrons. These are stopped in a surrounding blanket. The neutrons react with ^{6}Li to produce tritium for supplying the fusion region; energy is deposited in the blanket, from which heat is removed by a gas or liquid coolant.

be induced by powerful laser beams, by intense electron beams, and possibly by other means. In principle, the pellet containing the material for fusion may also contain fissile material, but the fission-fusion combinations of interest in the next section ordinarily use the neutrons from fusion to breed material in a separate blanket.

First-generation fusion devices would require a substantial supply of tritium. This may be produced by using neutrons from the D-T reaction to produce tritium from lithium 6:

$$^6\text{Li} + n \longrightarrow {}^4\text{He} + T + 4.6\,\text{MeV}$$

A basic D-T fusion device would have the schematic design shown in Figure 15-1. Energetic neutrons escape from a central reaction region. They are stopped in a cooled breeder blanket surrounding the fusion region. The blanket contains ^{6}Li, which yields tritium that is collected and fed, along with deuterium, to the fusion region. In principle, the central region could use either magnetic or inertial confinement. The heat that is delivered to the blanket by the neutrons is extracted to produce electricity (and run the fusion device).

Fusion devices would be even more technically advanced than fission reactors. Because of the difficulty of achieving break-even, i.e., between energy extracted from the blanket and energy supplied to the device, some consideration has been given to joining fission and fusion systems.

FISSION—FUSION

Fusion-produced neutrons could also be used to convert fertile materials, ^{238}U and ^{232}Th, to fissile material, principally ^{239}Pu and ^{233}U. Using a neutron to produce a nuclide which, at fission, releases 200 MeV may be regarded as more valuable than using it to produce a tritium nucleus. On the other hand, sufficient tritium must be available to supply the fusion reaction.

This imbalance could be alleviated, in principle, in a number of ways. The neutrons from fusion are much more energetic than those from fission. A D-T reaction liberates 17.6 MeV of energy, and the neutron carries off four-fifths of this or 14.1 MeV (the rest would be deposited in the fusion plasma). Neutrons of this energy are very effective in converting fertile material. Each one is capable of producing three or four fissile nuclei (the number from ^{238}U is somewhat greater than that from ^{232}Th), a sufficiently large number that, when they are eventually used in a fission device, neutrons could be used to produce tritium for return to the fusion reactor. On balance, a significant gain would have been made.

What is especially interesting is that this may all be accomplished in the blanket surrounding the fusion region. The fissile material being produced in the blanket would eventually start contributing energy from fissions induced by the fusion neutrons. This would add to the energy already available from the fusion neutrons and their interactions with fertile nuclei. The fissions also add more neutrons to the system. If, in addition to the fissionable material, lithium (containing some ^{6}Li) is in the blanket, the blanket can produce tritium, fissile material, and substantial amounts of energy.

In the equilibrium state, the blanket would be a subcritical fission reactor, using low-enrichment nuclear materials and some ^{6}Li. The blanket would have a coolant running through it, much as a fission core is cooled. However, the system would never reach criticality; rather, it would be fed energetic neutrons from the fusion reactors. From each fusion neutron, it seems possible to obtain, not only a replacement tritium atom, but hundreds of MeV of energy and a substantial fissile production. The basic conversion reactions are indicated in Figure 15-2.

Overall, such a hybrid system would be more complex than either a fusion device or fission reactor alone. The breeder blanket would contain lithium and tritium, as well as fissionable materials and fission products. The system would have many of the difficulties of both systems, the only exception being that, because the

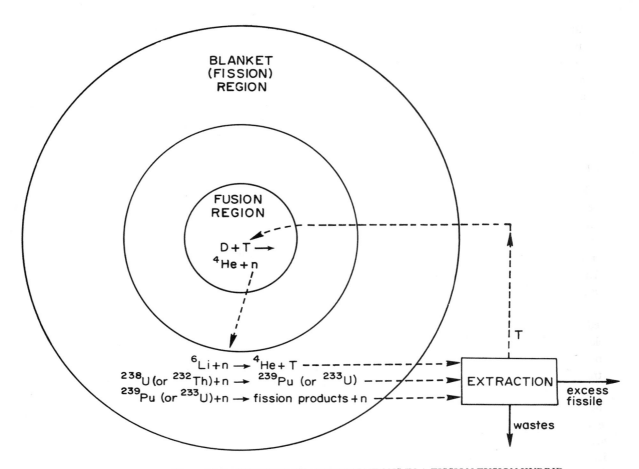

Figure 15-2. NUCLEAR TRANSFORMATIONS IN A FISSION-FUSION HYBRID.
The fusion region and the fission blanket region trade neutrons and tritium, much as in a pure
fusion reactor (Figure 15-1). In addition, the blanket itself acts as a fission breeder, receiving
fertile materials and discharging fissile material, fission products, etc.

fission assembly could be subcritical, reactivity control is not necessary. Because
criticality is not necessary, it may, in addition, be possible to design the system so
that the blanket avoids melting even if cooling is lost (and provided that the fusion
device turns off). In this way, the possibility of major releases of fission materials
and products could be avoided. On the other hand, the efficiency of the system
may be increased if the blanket is close to critical, in which case it would be
necessary to incorporate safety systems to enable the reactor blanket to withstand
loss-of-coolant accidents.

An alternative to a hybrid fission-fusion system, where the fusion blanket is
essentially a fission reactor, is to separate the systems substantially. The blanket
could be relatively pure fertile material, both ^{6}Li and ^{232}Th or ^{238}U, and bred
fissile would be used almost exclusively in separate fission reactors. There may be
economic or safety advantages to such a system. The design may also be easier than
that of a hybrid.

An indirect advantage to getting much more energy per fusion neutron than in a pure fusion device is that the inner wall of the fusion reactor may last much longer. One of the most formidable problems with fusion is to find a material to surround the fusion region that can withstand a large flux of 14 MeV neutrons. Because a hybrid device would effectively extract an order of magnitude more energy per neutron, this wall may have a proportionately longer lifetime.

ACCELERATOR BREEDERS

Simpler ways exist to produce high-energy particles than to use a fusion device. For many years researchers have used "accelerators" to produce beams of particles with energies from fractions of an MeV to tens of thousands of MeV. The latter are used by physicists investigating the basic structure of matter, but many lower energy devices have been adapted for routine uses in industry and even medicine. The possibility exists that a moderately high-energy device could be used to breed fissile material from fertile.

This approach would differ in one obvious respect from the breeding discussed earlier, whether in a conventional fission reactor or in a fusion blanket. The accelerator would use charged particles, such as protons, instead of neutrons from fusion or fission. The charged particles would strike a target to produce a shower of neutrons, which would then be used for fertile conversion.

Accelerators necessarily produce charged particle beams because they operate by using electrical or magnetic fields, or both, to accelerate particles, which must have an electrical charge to "feel" the fields. Many high-energy machines are linear devices which accelerate particles in a straight line by using intense microwave fields which the particles "ride" down the accelerator, picking up energy as they go. It is such a device, producing protons with energies of about 1000 MeV, that has recently aroused renewed interest for producing fissile material in a fertile assembly. Such an arrangement is indicated in Figure 15-3. This concept had been examined years ago, but was dropped because of the large amounts of energy necessary to run the accelerator. In recent years, however, designers have had much more experience

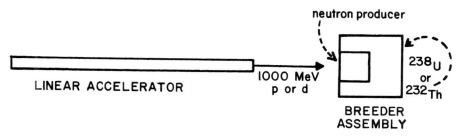

Figure 15-3. SCHEMATIC ACCELERATOR-BREEDER.
A linear accelerator supplies a beam of protons (although deuterons or other particles may be used). The beam strikes a primary target, producing a shower of neutrons, which convert fertile material to fissile in a breeder target.

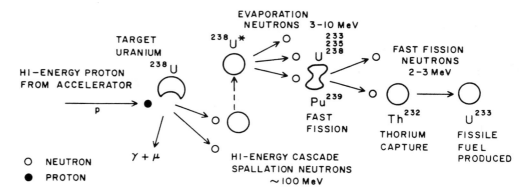

Figure 15-4. NUCLEAR REACTIONS IN AN ACCELERATOR BREEDER.
Neutrons are produced by a sequence of reactions: proton-induced spallation yields high-energy neutrons (about 100 MeV); these produce a larger number of evaporation neutrons, of several MeV; and these may produce fast fission neutrons. The total of about 50 neutrons per 1000 MeV proton can then produce fissile material. (Figure reproduced from BNL-50592.)

with high-current high-energy machines; as a result, the concept may be more practical, particularly when natural fissile resources are being depleted rapidly.

A variety of targets may be used for producing the neutron shower, but uranium itself is much more suitable than lighter materials. A 1000 MeV proton is estimated to produce about 50 neutrons on striking a ^{238}U target. In addition, the amount of energy released in this process (about 4000 MeV) is comparable to that required to operate the accelerator. The 50 neutrons would be used to produce fissile material in a breeding assembly. In many respects, this would be similar to a fast breeder blanket, except of course that no critical assembly would be involved. The precise composition of the target and breeder assembly would depend on a detailed design effort. It may, for example, be necessary to include some fissile material in the proton target in order to generate more neutrons and, more to the point, enough energy to run the accelerator. Neutron production and conversion mechanisms are indicated in Figure 15-4.

Such an accelerator-breeder could produce moderately large amounts of fissile material. Devices that are thought to cost about $1 billion (roughly the cost of a nuclear power plant) may be able to produce about 1000 kg of fissile material per year. As a rough estimate, this would correspond to a fissile cost of $100,000/kg (assuming the accelerator lasts significantly longer than ten years). This one metric ton of fissile would supply the annual feed requirements of only a few LWRs. However, with a system of high-conversion thermal reactors, where the annual feed is very small (say about 50 kg/year, with a conversion ratio of about 0.9), twenty reactors could be supplied. The cost of one accelerator breeder may not be excessive if it supplies many reactors.

This possibility is clearly of interest only when conserving uranium resources is an important consideration. One might think of the accelerator breeder, which is touted as only a moderate extension of current technology, as a possible backup to nuclear power systems that are not based on fast breeder reactors. In case thermal systems deplete uranium resources more rapidly than anticipated, the accelerator breeder could be called to duty.

NEAR-TERM PROSPECTS

It is extremely unlikely that either of the systems discussed in this chapter would be used during this century. The fusion program is proceeding rapidly, but several development stages remain before any fusion device, including a hybrid system, could be commercialized. The use of accelerators to breed fissile material, on the other hand, may be regarded as a useful approach only when uranium resources are sorely pressed, uranium prices are much higher than they are today, and alternative means of supplying reactors have failed or been exhausted.

It is, in fact, clear that for the rest of this century water-cooled reactors will constitute the bulk of nuclear electrical generating capacity in the United States and even abroad. However, considering the possible exhaustion of uranium resources, advanced reactors will constitute an increasing large share of new nuclear power plants. This is already becoming evident in other countries. This direction is not so clear in the United States, where the choice of energy supplies is greater and the potential for increasing the efficiency of energy use is substantial.

The possibilities for more resource-efficient plants are numerous, ranging from incorporation of new fuels in present light-water reactors to construction of entirely new reactor types. The most resource-efficient systems are those that breed or nearly breed and, as discussed in Chapter 10, these are essentially equivalent in a slowly growing energy economy. In any case, the reactor concepts likely to achieve any importance by the year 2000 are those described in Chapters 13 and 14: liquid metal and gas-cooled fast-breeder reactors; thorium-fueled light-water reactors, heavy-water reactors, and gas-cooled reactors; and even the molten salt breeder reactor.

The implications of relying on advanced systems are several. They are more efficient in their use of resources. But they often depend heavily on the availability of fuel recycle facilities. This carries with it the possibility of relatively enriched material reaching the hands of those who would make weapons. The decision whether to reprocess and recycle is therefore not an easy one, even though, in many countries, the choice has already been made.

It is not possible to predict what the energy picture will look like one or two decades from now. The choices to be made between now and then are not purely technical, but environmental, economic, social, and political. Only a few of these considerations, as they pertain to nuclear power, have been touched on in this book. However, several questions that are integral to the present status and the future growth of nuclear power have been set forth, as have the nuclear reactors from which nuclear energy is and will be derived.

Bibliography — Chapter Fifteen

"NTIS" indicates report is available from: National Technical Information Service, U.S. Department of Commerce, 5285 Port Royal Road, Springfield, VA 22161.

BNL-50592. "Linear Accelerator-Breeder, a Preliminary Analysis and Proposal," Brookhaven National Laboratory report BNL-50592 (November 1976) (NTIS).
> Consideration of the possibility of generating fissile material with a linear accelerator.

Emmett, J. L., Nuckolls, J., and Wood, L., "Fusion Power by Laser Implosion," *Scientific American,* vol. 230, p. 24 (June 1974).
> Describes how useful energy might be extracted from laser-induced fusion.

ERDA-4. "DCTR Fusion-Fission Energy Systems Review Meeting," Germantown, Md., December 3-4, 1974, U.S. ERDA report ERDA-4, U.S. Government Printing Office (1975).
> Collection of papers on various aspects of fission-fusion systems.

Metz, W. D., "Fusion Research (I): What is the Program Buying the Country?" *Science,* vol. 192, p. 1320 (1976); "Fusion Research (II): Detailed Reactor Studies Identify More Problems," *Science,* vol. 193, p. 38 (1976); "Fusion Research (III): New Interest in Fusion-Assisted Breeders," *Science,* vol. 193, p. 307 (1976).
> Series on the U.S. fusion program, including the possibility of fission-fusion hybrids.

Penner, S. S., *Nuclear Energy and Energy Policies,* vol. 3 of *Energy* (Addison-Wesley, Reading, Mass., 1976).
> Includes a description of fusion systems.

Reference Matter

APPENDIX A

Abbreviations and Units

AEC — Atomic Energy Commission

AECL — Atomic Energy of Canada, Limited

AGR — Advanced gas-cooled reactor

APS — American Physical Society

BWR — boiling-water reactor

CANDU — Canadian deuterium-uranium reactor

CFR — Code of Federal Regulations

CRBR — Clinch River Breeder Reactor

DOE — Department of Energy

ECC(S) — emergency core cooling (systems)

EPA — Environmental Protection Agency

EPRI — Electric Power Research Institute

ERDA — Energy Research and Development Administration

FBR — fast breeder reactor

FFTF — Fast Flux Test Facility

GCFR — gas-cooled fast (breeder) reactor

HPIS — high pressure injection system

HTGR — high-temperature gas-cooled reactor

HWR — heavy-water reactor

IAEA — International Atomic Energy Agency

LMFBR — liquid-metal fast breeder reactor

LOCA — loss-of-coolant accident

LPIS — low pressure injection system

LPZ — low population zone

LWBR — light-water breeder reactor

LWR — light-water reactor

MSBR — molten salt breeder reactor

NRC — Nuclear Regulatory Commission

NSSS — nuclear steam supply system

PCRV — prestressed concrete reactor vessel

PWR — pressurized-water reactor

RHR — residual heat removal (system)

SAR — safety analysis report

SSCR — spectral shift control reactor

245

Table of the Elements

Element	Symbol	Atomic Number	Mass Number of Principal Isotopes[a] (percent natural abundance)
hydrogen	H[b]	1	1 (99.98), 2 (0.02)
helium	He	2	4 (100)
lithium	Li	3	7 (93), 6 (7)
beryllium	Be	4	9 (100)
boron	B	5	11 (80), 10 (20)
carbon	C	6	12 (98.9), 13 (1.1)
nitrogen	N	7	14 (99.6), 15 (0.4)
oxygen	O	8	16 (99.8), 18 (0.2)
fluorine	F	9	19 (100)
neon	Ne	10	20 (91), 22 (9)
sodium	Na	11	23 (100)
magnesium	Mg	12	24 (79), 26 (11)
aluminum	Al	13	27 (100)
silicon	Si	14	28 (92), 29 (5)
phosphorus	P	15	31 (100)
sulfur	S	16	32 (95), 34 (4)
chlorine	Cl	17	35 (76), 37 (24)
argon	Ar	18	40 (99.6), 36 (0.3)
potassium	K	19	39 (93), 41 (7)
calcium	Ca	20	40 (97), 44 (2)
scandium	Sc	21	45 (100)
titanium	Ti	22	48 (74), 46 (8)
vanadium	V	23	51 (99.8), 50 (0.2)
chromium	Cr	24	52 (84), 53 (10)
manganese	Mn	25	55 (100)
iron	Fe	26	56 (92), 54 (6)
cobalt	Co	27	59 (100)
nickel	Ni	28	58 (68), 60 (26)
copper	Cu	29	63 (69), 65 (31)
zinc	Zn	30	64 (49), 66 (28)
gallium	Ga	31	69 (60), 71 (40)
germanium	Ge	32	74 (36), 72 (27)
arsenic	As	33	75 (100)
selenium	Se	34	80 (50), 78 (24)
bromine	Br	35	79 (51), 81 (49)
krypton	Kr	36	84 (57), 86 (17)
rubidium	Rb	37	85 (72), 87 (28)
strontium	Sr	38	88 (83), 86 (10)
yttrium	Y	39	89 (100)
zirconium	Zr	40	90 (51), 94 (17)
niobium	Nb	41	93 (100)
molybdenum	Mo	42	98 (24), 96 (17)
technetium	Tc	43	99[c]
ruthenium	Ru	44	102 (32), 104 (19)
rhodium	Rh	45	103 (100)
palladium	Pd	46	106 (27), 108 (27)
silver	Ag	47	107 (52.4), 109 (48)
cadmium	Cd	48	114 (29), 112 (24)
indium	In	49	115 (96), 113 (4)
tin	Sn	50	120 (33), 118 (24)
antimony	Sb	51	121 (57), 123 (43)
tellurium	Te	52	130 (34), 128 (32)
iodine	I	53	127 (100)
xenon	Xe	54	132 (30), 129 (26)
cesium	Cs	55	133 (100)
barium	Ba	56	138 (72), 137 (11)
lanthanum	La	57	139 (99.9), 138 (0.1)
cerium	Ce	58	140 (88), 142 (11)

Table of the Elements (Continued)

Element	Symbol	Atomic Number	Mass Number of Principal Isotopes[a] (percent natural abundance)
praeseodymium	Pr	59	141 (100)
neodymium	Nd	60	142 (27), 144 (24)
promethium	Pm	61	145[c]
samarium	Sm	62	152 (27), 154 (23)
europium	Eu	63	153 (52), 151 (48)
gadolinium	Gd	64	158 (25), 160 (22)
terbium	Tb	65	159 (100)
dysprosium	Dy	66	164 (28), 162 (26)
holmium	Ho	67	165 (100)
erbium	Er	68	166 (33), 168 (27)
thulium	Tm	69	169 (100)
ytterbium	Yb	70	174 (33), 172 (22)
lutetium	Lu	71	175 (97), 176 (3)
hafnium	Hf	72	180 (35), 178 (27)
tantalum	Ta	73	181 (99.99), 180 (0.01)
tungsten	W	74	184 (31), 186 (28)
rhenium	Re	75	187 (62), 185 (38)
osmium	Os	76	192 (40), 190 (26)
iridium	Ir	77	193 (63), 191 (37)
platinum	Pt	78	195 (34), 194 (33)
gold	Au	79	197 (100)
mercury	Hg	80	202 (30), 200 (23)
thallium	Tl	81	205 (70), 203 (30)
lead	Pb	82	208 (52), 206 (24)
bismuth	Bi	83	209 (100)
polonium	Po	84	210[c]
astatine	At	85	215[c]
radon	Rn	86	222[c]
francium	Fr	87	223[c]
radium	Ra	88	226 (100)
actinium	Ac	89	227[c]
thorium	Th	90	232 (100)
protactinium	Pa	91	231 (100)
uranium	U	92	238 (99.3), 235 (0.7)
neptunium	Np	93	237[c]
plutonium	Pu	94	239[c]
americium	Am	95	241[c]
curium	Cm	96	244[c]
berkelium	Bk	97	249[c]
californium	Cf	98	252[c]
einsteinium	Es	99	253[c]
fermium	Fm	100	255[c]
mendelevium	Md	101	256[c]
nobelium	No	102	254[c]
lawrencium	Lr	103	256[c]
		104	257[c]
		105	260[c]

a. Where more than one isotope occurs naturally, the two with greatest abundance are given. For more detailed information, see C. M. Lederer et al., Table of Isotopes, 6th ed. (Wiley, New York, 1967); or Handbook of Chemistry and Physics, 58th ed. (CRC Press, Cleveland, 1977).

b. The mass 2 and 3 isotopes of hydrogen have their own names, i.e., deuterium (D) and tritium (T), respectively.

c. For short-lived elements or those that do not occur naturally, one of the more important isotopes is given.

MISCELLANEOUS UNITS AND EQUIVALENCES

SCIENTIFIC NUMERICAL NOTATION

A number $\times 10^n$ means "multiply the number by 10 n times (or divide if n is a negative number)"

BASIC UNITS IN THE METRIC SYSTEM

Quantity	Unit (abbreviation)
mass	gram (gm)
length	meter (m)
time	second (sec)
force	newton (n) = kg-n/sec^2
pressure	pascal (Pa) = n/m^2
energy	joule (j) = n-m
power	watt (W) = j/sec

Multiples of any basic unit may be formed by using the following prefixes:

Multiple	Prefix (abbreviation)
10^9 (billion)	giga (G)
10^6 (million)	mega (M)
10^3 (thousand)	kilo (k)
10^{-3} (thousandth)	milli (m)
10^{-6} (millionth)	micro (μ)
10^{-9} (billionth)	nano (n)

ENERGY-RELATED

1 MeV = 1 million electron volts (eV) = 1.52×10^{-16} Btu = 1.60×10^{-13} watt-sec (or joule)

1 Btu = 1 British thermal unit = 1.05×10^3 joule

1 erg = 10^{-5} joule

1 kilowatt-hour (kWh) = 3413 Btu = 3.6×10^6 joule

1 megawatt-day (MWd) = 8.2×10^7 Btu = 8.6×10^{10} joule

Typical fission energy yield = 200 MeV = 3.0×10^{-14} Btu = 0.1% of ^{235}U mass*

1 gigawatt = 1000 megawatts; 1 megawatt = 1000 kilowatts; 1 kilowatt = 1000 watts

1 watt = 1 joule/sec

MASS-RELATED

1 atomic mass unit (amu) = $1/12 \times {}^{12}$C mass = 1.66×10^{-24} gm = 931.44 MeV*

neutron mass = 1.0087 amu = 1.67×10^{-24} gm = 939.5 MeV*

proton mass = 1.0073 amu = 1.67×10^{-24} gm = 938.2 MeV*

electron (beta particle) mass = 1/1836 × proton mass = 0.511 MeV*

^{235}U mass = 235.04 amu

1 metric ton (Mg or Te) = 2205 lb = 1000 kg

1 pound (lb) = 454 gm = 0.454 kg

1 ton = 2000 lb = 907 kg = 0.907 Mg

*Mass-energy equivalence.

TEMPERATURE EQUIVALENCE

(Centigrade temperature: C)
Kelvin temperature: $K = C + 273$
Farenheit temperature: $F = 9/5C + 32$
Rankine temperature: $R = 9/5C + 492 = F + 460$

RADIATION

1 Curie (Ci) = 3.7×10^{10} disintegrations per second
1 rad = 0.01 joule/kg (energy deposited) = 1000 mrad
dose equivalent in rem = dose in rad $\times$ relative biological effectiveness
1 rem = 1000 mrem

OTHER

1 centimeter (cm) = 0.01 m
1 inch (in) = 2.54 cm = 0.0254 m
1 foot (ft) = 30.48 cm = 0.3048 m
1 barn = 10^{-24} cm^2
1 cubic foot = 28.3 liter = 2.83×10^4 cm^3 = 0.0283 m^3
1 gallon = 3.78 liter
atmospheric pressure $\simeq$ 14.7 psi = 14.7 lb/in^2
1 psi = 1 lb/in^2 = 0.0069 MPa

APPENDIX B

Reactions, Cross-Sections, and Moderation

The basic process occurring in a nuclear reactor, maintenance of the chain reaction, is nuclear, as opposed to "atomic" or "chemical." It depends on interactions between atomic nuclei or "parts" thereof for its continuation. The energy made available from fission events is due to the differences in nuclear potential energy (or, let us say, mass) of the initial and final particles. Moreover, in present-day reactors, the basic initiator of fission events, the neutron, is slowed down by nuclear collisions, thereby increasing its probability of inducing fission.

The fundamental physical law to which a nuclear reactor "owes" the possibility of its existence is the conservation of energy. This must be thought of in its most general terms, i.e., one must include the fact that an amount of mass has an energy equivalence, $E = mc^2$, where c is the speed of light. Before this century, it was well understood that in a collision between two or more particles, if the outgoing particles differed from the incoming, energy conservation still worked, provided one took account of the modification of the *internal* energies of the particles. Only in this century has it been recognized that this modification manifests itself as a change in *mass*. The result is that energy conservation may often be seen as the statement that the change in the kinetic energy of the system is balanced by its change in mass. As an example, take the following reaction:

$$^{235}U + \text{thermal neutron} \rightarrow 2 \text{ fission fragments} + 2 \text{ neutrons} + 200 \text{ MeV}$$

(where the 200 MeV could manifest itself as gamma rays, neutrinos, kinetic energy, and, perhaps, production of some electrons). Energy conservation is implicit in this as: the mass-energy of the initial particles equals that of the final particles + 200 MeV.

Granting energy conservation, much more could be said about a reaction such as this. Other quantities, such as linear momentum and angular momentum, would be conserved. More importantly for our purposes, we can describe in some precise way the *probability* of various outcomes of some initial condition.

Consider an interaction between two particles and assume, for the moment, that both particles survive the "collision," but that their state of motion may be altered. The total probability of such interaction is measured by the elastic cross-section σ_e, where "elastic" means the particles themselves are not changed. (Historically, it meant that no kinetic energy was lost in the collision, but as discussed

250

above, this must be true if the particles are unchanged and no new particles are produced.) Such a scattering event may be specified in more detail. One basic specification is the direction in which one of the particles is scattered, relative to its original direction. This angular dependence must be considered in detail to arrive at some of the results on neutron moderation given below.

The normalization for cross-section is a natural one, consistent with its interpretation as an effective collision area. Consider an artificial case in which we have N targets, each with an area σ, and a pointlike projectile which interacts with a target if (and only if) it strikes one of those areas σ. If we direct our projectile at an area A throughout which the N targets are dispersed, the probability of interaction is clearly the ratio of Nσ to A. For more general kinds of interactions, the "cross-section" σ is defined so that the interaction probability is still Nσ/A. (If, then, we have a *current* of projectiles i, the *rate* at which interactions occur is $r = i\sigma N/A$.) This same definition also applies to possible outcomes of an interaction other than elastic scattering, such as:

(1) The two interacting objects may effectively "stick" together as in neutron capture, where a nucleus absorbs a neutron, giving off a gamma ray in the process. The cross-section corresponding to capture is σ_c.
(2) Two or more different particles may emerge from the interaction. The most important example of this, for reactors, is fission, where the cross-section is labeled as σ_f. Another possibility in this class would be n + nucleus → n + n + nucleus$'$.

For a single initial state, the sum of cross-sections for all outcomes other than elastic scattering is the absorption cross-section, which for a neutron-nucleus interaction is made up of many parts, including the two given above: $\sigma_a = \sigma_f + \sigma_c + \ldots$ The balance between σ_f and σ_c is an important matter for any nuclear fuel (see Appendix C).

These cross-sections have an important dependence on energy. For example, the cross-section for fission of ^{235}U is given schematically in Figure B-1. The relatively large cross-section at low energy is a reason for "moderating" neutrons down from the MeV-range energies they have on emergence from fission events.

This moderation is achieved by including (as is done in "thermal" reactors) a large amount of low-mass material with which neutrons can collide, thereby reducing their energy. The most effective material for accomplishing this is hydrogen, because its nucleus (one proton) has essentially the same mass as a neutron. It can, therefore, accept a large portion of the neutron energy as recoil, whereas a heavy (that is to say, massive) nucleus can accept only a small amount of energy in an elastic collision. This can be seen by considering head-on elastic collisions (essentially "billiard-ball" collisions) for two types of target. If the target has the same mass as the projectile neutron, in a classical head-on collision the neutron will stop, giving up all its energy to the target, which then proceeds with the same velocity the neutron originally had. At the other extreme, if the target is much heavier than the neutron, the neutron simply bounces backward, and its new energy is only slightly less than its old. Although the exact details depend on the neutron energy and angular considerations, it can be shown that to thermalize neutrons many more collisions are necessary with heavy targets than with light. To be more precise, we may express the result in terms of logarithms, i.e., that the *average decrease* (per collision) *in the natural log of the neutron energy* is,

$$\xi = (\ln E_{initial} - \ln E_{final})_{ave} \simeq 1 + \frac{(A-1)^2}{2A} \ln \frac{A-1}{A+1} \ ,$$

where A is the nucleon number of the target nucleus. For A = 1 (a hydrogen nucleus, i.e., a proton), ξ is actually 1; for A > 1, it is approximately 2/(A + 2/3). In an average (not necessarily head-on) collision with a proton, a neutron would be

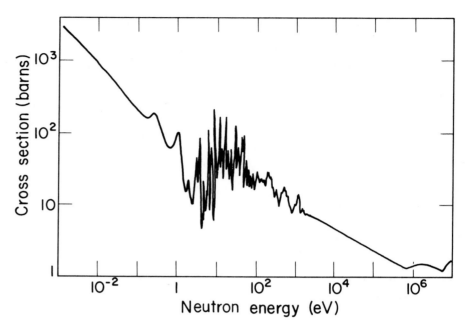

Figure B-1. SCHEMATIC FISSION CROSS-SECTION FOR ^{235}U.
The overall dependence of ^{235}U fission cross-section on neutron energy is moderately smooth, except for the region where resonances cause the cross-section to vary rapidly with energy. (Figure courtesy of Lawrence Berkeley Laboratory.)

reduced to 1/2.72th of its initial energy. More to the point, to reduce the neutron energy the eight orders of magnitude from, say, 2 MeV (the typical energy of a fission neutron) to thermal energy (1/40 eV) requires

$$\frac{\ln(2 \times 10^6) - \ln(1/40)}{1} = 18$$

collisions with protons. The corresponding numbers for some other interesting elements are deuterium 25, helium 43, beryllium 86, carbon 114, uranium 2172.

A more useful measure of the effectiveness of a moderating material would fold in the cross-section for the collision and the number density of the scattering centers in the material. The "slowing down power" is such a quantity and it is

$$S = \sum_i N_i \sigma_{ie} \xi_i \ ,$$

where i corresponds to the ith type of atom in the moderating material, N is the number per unit volume, σ_e the elastic cross-section, and ξ the average logarithmic decrease. S is an average logarithmic decrease *per unit length* that the neutron traverses.

The slowing down power still does not take account of the possibility of *absorption* of the neutron by the moderator; a moderator that robs the system of many neutrons can be a great disadvantage. A useful quantity is the "moderating ratio"

$$S/\sum_i N_i \sigma_{ia} \ ,$$

where σ_{ia} is the absorption cross-section of the ith constituent. In a sense, this is the ratio of the slowing down power to the absorptive power.

TABLE B-1.
Slowing Down Properties of Moderators

Moderator	Slowing Down Power (cm^{-1})	Moderating Ratio
Water	1.28	58
Heavy water	0.18	21,000
Helium[a]	10^{-5}	45
Beryllium	0.16	130
Graphite	0.065	200

a. At atmospheric pressure and temperature.

Source: S. Glasstone and A. Sesonske, *Nuclear Reactor Engineering* (Van Nostrand Reinhold, New York, 1963), reproduced by permission.

These last two quantities are strongly energy dependent, since they involve cross-sections which depend on energy. However, suitably weighting them, they can be useful. Some results given in Table B-1 are illuminating. There it is seen that heavy water, because it rarely absorbs neutrons, has a much larger moderating ratio than other materials. (This is one reason that the CANDU can sustain a chain reaction for long periods of time with only natural uranium as a fuel.)

APPENDIX C

Characteristics of Nuclear Fuel Materials

The essential constituent of a nuclear reactor is the fissionable material contained in its core. The energy available from the reactor is basically determined by how many fissions[1] occur in the course of the chain reaction; continuation of the chain reaction depends on the neutrons which become available, principally as a result of fission.

Of the fissionable nuclides in Table C-1, those whose neutron-induced fission threshold is zero (^{235}U, ^{239}Pu, ^{233}U) are referred to as "fissile," while the others are "fertile," since neutron capture onto them can lead, via the conversion sequences indicated there, to a fissile nuclide. However, even the fertile nuclides in the table are subject to "fast fission," i.e., fission induced by neutrons of energy above the indicated nonzero threshold. The "fissile" materials are those that can serve as the source for most of the fissions in a reactor where the neutrons produced are very rapidly "thermalized" to low energies (see Appendix B).

All the above materials are radioactive. For the fissionable nuclides listed, this is not relevant to their behavior in a reactor since their half-lives are so long (see Table C-2), but can affect externalities, such as the danger they present to humans. However, the radioactive character of intermediate nuclides in a conversion sequence can be of great importance, since they lead to more useful (i.e., fissile) nuclides. For example, protactinium 233, one of the intermediate stages in the thorium cycle conversion, has so long a half-life that the probability of its capturing a neutron before decaying to ^{233}U may cause some difficulty in achieving high conversion ratios. Table C-2 contains information on the most important nuclides occurring in reactors.

Having noted that the neutrons generated by fission can either produce more fissions (by interacting with a fissionable material) or produce fissile material (by being captured by a fertile material), we present some basic information about neutron production, and fission and capture probabilities. Table C-3 gives the cross-section for thermal (read "slow") neutron-induced fission and capture on the indicated nuclides. For the fissile materials, ^{233}U, ^{235}U, ^{239}Pu, a small capture/fission ratio is an advantage because neutrons captured onto them are lost. The fact that ^{232}Th has a higher capture cross-section than ^{238}U accounts in part for a higher

[1] At an average energy of about 200 MeV per fission (or 77 million Btu or 22 thousand kWh [thermal] per gram of fissioned ^{235}U). Most of the energy is released as kinetic energy of the fission fragments, which typically stop in the fuel itself, thus heating up the fuel material. Of the energy carried off by other particles or radiation ($n,\alpha,\beta,\gamma,\nu$) due to fissions or subsequent decays, only the energy associated with the neutrinos, a few MeV, on the average, escapes from the reactor.

254

TABLE C-1.
Principal Fissionable Nuclides

Nuclide	Fission Threshold (Neutron Energy in MeV)	Average Energy Available from Thermal Fission (MeV)
Fertile ^{232}Th	1.4	—
Fissile ^{233}U	0	198
Fissile ^{235}U	0	202
Fertile ^{238}U	0.6	—
Fissile ^{239}Pu	0	210

Conversion Sequences (half-lives are given in Table B-2):

"Uranium-Plutonium Cycle" $^{238}\text{U} + \text{n} \xrightarrow{\gamma} {}^{239}\text{U} \xrightarrow{\beta} {}^{239}\text{Np} \xrightarrow{\beta} {}^{239}\text{Pu}$

"Thorium-Uranium Cycle" $^{232}\text{Th} + \text{n} \xrightarrow{\gamma} {}^{233}\text{Th} \xrightarrow{\beta} {}^{233}\text{Pa} \xrightarrow{\beta} {}^{233}\text{U}$

TABLE C-2.
Radioactivity of Nuclear Materials

Nuclide	Activity	Half-Life[a]
^{232}Th[b]	α	1.41×10^{10} yr
^{238}U[b]	α	4.51×10^{9} yr
^{235}U[b]	α	7.1×10^{8} yr
^{233}Th	β	22.2 min
^{233}Pa	β	27.0 days
^{233}U	α	1.62×10^{5} yr
^{239}U	β	23.5 min
^{239}Np	β	2.35 days
^{239}Pu	α	2.44×10^{4} yr
^{232}U	α [+ γ rays]	72 yr

a. Lederer et al., *Table of Isotopes,* 6th ed., (Wiley, New York, 1967).
b. Occur in substantial quantities in nature.

TABLE C-3.
Thermal Neutron Cross-Sections for Nuclear Materials[a]

Nuclide	Fission (barns)	Capture (barns)	$\dfrac{\text{Capture}}{\text{Fission}}$ ratio
^{232}Th	—	7.4	—
^{233}U	527	54	0.102
^{235}U	577	106	0.184
^{238}U	—	2.7	—
^{239}Pu	742	287	0.387

a. See Appendix B for discussion of cross-section.

TABLE C-4.
Neutrons Liberated per Neutron Capture in Fissile Materials

Nuclide	ν, Per Thermal Neutron Induced Fission	η, Per Thermal Neutron Absorbed	η, Per Fast Neutron Absorbed
233U	2.50	2.27	2.60
235U	2.43	2.06	2.18
239Pu	2.90	2.10	2.74

conversion ratio in the thorium than in the plutonium cycle because of the greater probability that the thorium cycle conversion sequence (see above) is initiated.

Another primary consideration is the number of neutrons produced per fission, ν, given as the first column of Table C-4. These neutrons are the currency of the reactor's economy. However, since, in the neutron economy, we want to know what return we get on a neutron absorbed (including those lost to capture), we can correct ν, using the capture/fission ratio, to obtain η, the number of neutrons liberated per neutron absorbed. We also give the equivalent number for a fast (unmoderated) neutron spectrum. For *thermal* reactors, of these fissile nuclides,

Figure C-1. NEUTRON YIELD VERSUS ENERGY OF ABSORBED NEUTRON FOR FISSILE MATERIALS.
The figure shows the dependence of neutron yield per neutron absorbed (eta, η) versus energy of absorbed neutron for the three principal fissile isotopes, 235U, 239Pu, and 233U. For substantial energy ranges, the neutron yield is equal to or greater than 2, the minimum value for producing as much fissile material as is consumed. (Figure reproduced from ERDA-1541.)

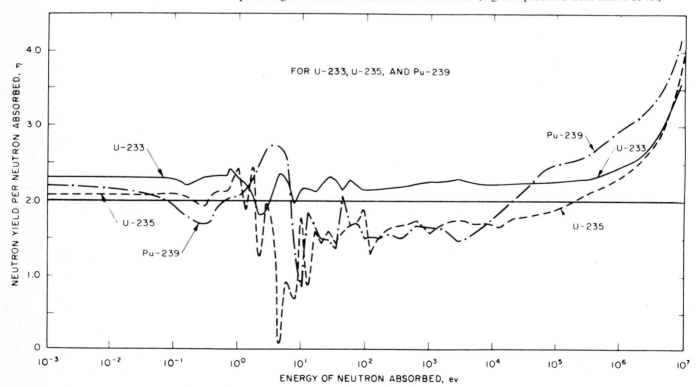

^{233}U liberates the most neutrons per neutron absorption. Since thorium is also relatively easily converted to ^{233}U, this accounts roughly for the advantage of the thorium cycle over the plutonium.

For the fast breeder reactor, ^{235}U is clearly the poor cousin of either ^{233}U or ^{239}Pu, since it yields just barely more than two neutrons per neutron absorbed, and a greater excess is required, considering losses, if we are to continue the chain reaction (one neutron), produce a new fissile nucleus (a second neutron), and have an excess left over to breed *extra* fissile materials.

Actually, although the numbers in Table C-4 do indicate qualitatively the differences between fissile fuels, they are not enough to yield conversion ratios because

(1) They only indicate results from fission due to thermal and fast neutrons, neglecting neutrons of intermediate energy. Thus, although the thermal values for η given above are roughly correct for E_n = 0.01 to 1.0 eV for ^{233}U and ^{235}U, η is noticeably less than 2.0 for most of this region in the case of ^{239}Pu.[2]

(2) They do not account for the relative importance of neutron capture by fertile material as opposed to the reactor structure or coolant.

A more complete representation of the behavior of η with energy is given in Figure C-1. Although we have spoken only of "thermal" reactors and "fast" reactors, ^{233}U has a behavior that would actually enable it to function well at intermediate energies.

[2] Thermal as used here is 1/40 eV.

APPENDIX D

Thermal Efficiency and Cooling

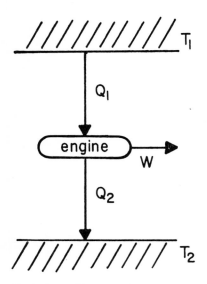

Figure D-1. SCHEMATIC HEAT ENGINE.

An amount of heat Q_1 is supplied to the engine from a reservoir at temperature T_1. Of the total supplied, W is converted to work and Q_2 is rejected to the reservoir with lower temperature T_2.

Although a large light-water reactor generates heat at the rate of about 3000 megawatts, the electrical output of the plant is only 1000 MWe, representing a conversion efficiency of 33%. The fact that this is so much less than 1 reflects primarily a basic limitation to the efficiency of "heat engines," devices that convert thermal energy to mechanical energy.

A heat engine can be idealized as withdrawing heat from a large reservoir at a fixed temperature T_1 and, operating in cyclic fashion, converting a portion of that energy to work, while rejecting the remainder to a second large reservoir at a temperature T_2, lower than T_1 (see Figure D-1). For a device operating over a range of temperatures, a combination of these idealized devices may be used to represent the system. The efficiency of our engine is given by the ratio W/Q_1, where Q_1 is the amount of heat withdrawn from the high-temperature reservoir and W is the work performed by the engine. By conservation of energy, $W = Q_1 - Q_2$, where Q_2 is the heat rejected to the low-temperature reservoir. These comments assume, clearly, that the internal energy of the heat engine itself does not change, at least from one cycle to the next. (Any cyclic device satisfies this requirement since, at the end of each cycle, it has returned to its state at the end of the previous cycle.)

It is a basic question in thermodynamics to inquire what is the best possible efficiency for an engine operating between temperatures T_1 and T_2. One is led to the answer by considering a "Carnot" engine, one that uses a working fluid on a cycle with four parts. Expressed in terms of a gas confined by a piston: the gas is allowed to absorb heat Q_1 at a constant temperature T_1 (the gas will expand, doing work on the piston); the gas is allowed to expand further, but with no heat transfer (i.e., adiabatically), until its temperature decreases to T_2; the gas then *gives up* heat Q_2 at a constant temperature T_2 until the volume decreases just enough; the gas is compressed adiabatically until its temperature rises to T_1 and it returns to its original volume. In the first two stages, the gas does work on the piston; in the second two, the piston does work on the gas. The difference is the net work $W = Q_1 - Q_2$ and it can be shown that for an "ideal" gas, $W = (1 - T_2/T_1) Q_1$.[1]

[1] For this result to hold, the temperatures T_1 and T_2 must be given on an "absolute" temperature scale, one where $0°$ specifies absolute zero. The absolute scale corresponding to the Celsius degree is the Kelvin scale (one on which water freezes at 273 °K); that corresponding to the Fahrenheit degree is the Rankine scale (0 °F = 460 °R).

258

It can further be shown that no heat engine can achieve a better efficiency than the Carnot result, $W/Q_1 = 1 - T_2/T_1$. Note that this efficiency approaches 1 only as T_2 approaches absolute zero or T_1 becomes very large. A reactor that is generating heat in a core whose temperature is 600 °F (1060 °R) and that ultimately rejects heat to a 70 °F (530 °R) river has a Carnot efficiency of $1 - 530/1060 = 0.5$. The only way — in principle — to improve this is to lower the temperature of the cooling river or raise the temperature at which the reactor operates. But on earth the rejection temperature can never be much lower than 500 °R (300 °K). And it is hard to imagine operating a water-cooled reactor at higher than 600 °F, where the pressures required are on the order of 2000 psi. The situation can be improved by using a gaseous coolant, or even metal, at higher temperature. Steam is not appropriate because at high temperatures it is very corrosive of core materials. Helium or CO_2 are the usual gaseous alternatives — see HTGR, Chapter 8.

There are two obvious reasons for which the Carnot efficiency is not achieved. One is the rather general fact that actual operating systems include turbulence, friction, and temperature differentials across system components that preclude adherence to any idealized cycle. A second is directly pertinent to systems that include the boiling of water, since the boiling and condensation that occur in a steam cycle are reasonably well described as constant pressure processes (isobars), rather than the constant temperature (isothermal) processes required for the Carnot cycle. The steam cycle, therefore, is usually approximated by the Rankine cycle, which is described as follows (in terms of water taken from the condenser after the turbine): the water from the condenser, which is at a relatively low pressure, is subjected to adiabatic compression to reach the pressure of the boiler; in the boiler, which operates at a relatively fixed pressure, the water is heated isobarically to produce steam, which is heated even above the boiling temperature; the steam is expanded adiabatically (in an expansion chamber, the turbine); and the low-pressure steam from the chamber undergoes isobaric isothermal condensation to form water which then begins the cycle again. This cycle cannot achieve the theoretical efficiency available for the Carnot cycle.

For both of the reasons above, the achievable efficiency is reduced below that suggested by the expression $1 - T_2/T_1$. On implementation, the efficiencies available in water-cooled reactor power plants are 30% to 33%, less than from fossil-fired generating plants, where the fuel is burned at higher temperatures. These efficiencies are often expressed in an alternative form, the "heat rate" (actually the reciprocal of the efficiency), which gives the amount of thermal energy required per electrical energy produced, usually in units of Btu/kWh. (One can then easily show that typical heat rates are about 9000 to 11,000 Btu/kWh; an efficiency of 1 would give 3413 Btu/kWh.)

APPENDIX E

Criticality and Control

Chapter 1, Appendix B, and Appendix C discussed some of the basic notions of reactor physics. However, the idea of criticality has only been suggested. It is clear that for maintenance of a given power level, an essentially steady-state condition, as many neutrons must be "born" as are absorbed. The relative size of the rates at which neutrons appear and are absorbed is expressed in terms of the "multiplication factor" k for the reactor; k may be defined as the ratio of the number of fissions in one generation to the number in the previous generation. (It may also be defined as the ratio of the number of neutrons in succeeding generations.) For a steady-state condition, k must equal 1, in which case the reactor is said to be "critical." A growing fission rate (a "super-critical" condition) is implied by $k > 1$, a decreasing rate by $k < 1$.

As discussed in Chapter 1, the short neutron "lifetime," order of 10^{-4} s in a thermal and even less in a fast reactor, would make reactors quite unstable were it not for the fact that a small fraction of the neutrons from any generation are "delayed." These neutrons are not produced directly in fission events. Instead, they occur when the decay of a fission product (or possibly a fission product daughter) leaves some nucleus in a state that is unstable to neutron decay. Though this neutron decay, if it occurs, will proceed promptly, it will have been delayed by the time it took the parent to beta decay. A reactor at steady power is subcritical on prompt neutrons, but just critical when the delayed neutrons are included. The speed of any change in the power level is determined essentially by the generation time of the delayed neutrons, provided the reactor remains subcritical on prompt neutrons. Conversely, if the reactor goes "prompt critical," the advantage offered by the presence of delayed neutrons is lost.

The decay of any radioactive species can be characterized by a "half-life" $t_{1/2}$, the time it takes for half of the nuclei in a sample to decay. (Alternatively, one can express the same information in terms of the decay constant: the rate of disintegrations in a sample will be proportional to the number of atoms remaining, N. Therefore $dN/dt = -\lambda N$, λ being the "decay constant." It is straightforward to show that if N_o atoms are present at time $t = 0$, the number at any time thereafter is $N = N_o e^{-\lambda t}$; and further $\lambda t_{1/2} = \ln 2 = 0.693$.) The time dependence of the appearance of delayed neutrons makes it clear that they have a number of different precursors (fission products leading to these neutrons) with a range of half-lives.

260

TABLE E-1.
Delayed Neutron Data for Thermal Fission in ^{233}U, ^{235}U, and ^{239}Pu

Group	^{233}U Half-Life (second)	^{233}U Yield (neutrons per fission)	^{235}U Half-Life (second)	^{235}U Yield (neutrons per fission)	^{239}Pu Half-Life (second)	^{239}Pu Yield (neutrons per fission)
1	55.00	0.00057	55.72	0.00052	54.28	0.00021
2	20.57	0.00197	22.72	0.00346	23.04	0.00182
3	5.00	0.00166	6.22	0.00310	5.60	0.00129
4	2.13	0.00184	2.30	0.00624	2.13	0.00199
5	0.615	0.00034	0.610	0.00182	0.618	0.00052
6	0.277	0.00022	0.230	0.00066	0.257	0.00027
Total yield:	–	0.0066	–	0.0158	–	0.0061
Total delayed fraction (β):		0.0026	–	0.0065	–	0.0021

Source: Adapted from Keepin, *Physics of Nuclear Kinetics*, © 1965, Addison-Wesley, Reading, Mass.

Historically, these neutrons have been fit into six relatively well-defined groups, corresponding to half-lives from 0.2 to 56 sec. Measurements of delayed neutrons from fission of various nuclides has shown, though, that a given group has a half-life that varies slightly from one nuclide to another. Table E-1 gives the measured half-lives of the six groups for thermal-neutron fission of ^{233}U, ^{235}U, and ^{239}Pu. The observed variation in half-life indicates that the groups are due to more than one precursor, with slightly differing half-lives, and that fission of each possible nuclide populates the precursors differently. Most, but not all, the precursors have been identified.

Table E-1 also gives the neutron yield for each group. The most important single number, for each nuclide, is the total delayed fraction β, the ratio of delayed neutron yield to total neutron yield from fission. For ^{235}U, β is 0.0065; it is smaller for ^{233}U and ^{239}Pu. For both uranium cases, the fraction is somewhat larger for fast-neutron-induced fission; ^{239}Pu is unchanged. Delayed neutrons can also arise from fertile materials. In the case of ^{239}Pu-fueled fast reactors, this contribution to the total delayed fraction is quite important, raising the total β significantly higher than the 0.2% from ^{239}Pu alone. In any case, the delayed-neutron fraction is sufficiently large to prevent the very rapid changes in neutron population suggested by the prompt neutron lifetime (see Chapter 1). As a result, rather slowly acting means of control, insertion of rods, addition of chemicals, etc., are sufficient to maintain a constant or slowly changing power level. This is particularly true since, in a power reactor, k is never allowed to be much different from 1.

In passing, we should note that there is a situation where the delayed neutrons would have no practical effect, and that is a condition of "prompt criticality." As is suggested by the name, this indicates that the multiplication factor exceeds 1 without counting the delayed neutrons. There is then no fundamental restraint on a rapid increase in the number of neutrons.

Although we have avoided the concept so far, "reactivity" is a useful term in this context. The reactivity ρ of a reactor is defined to be the ratio $(k-1)/k$ and is a quantity that occurs regularly in equations describing the neutron population. A reactor that is just critical, then, has $\rho = 0$. Prompt criticality occurs when $\rho = \beta$. It is, therefore, important that ρ be kept less than β. We will return to this question.

It is worth considering in slightly greater detail what contributes to the multiplication factor. Appendix C discusses one very important aspect of this question, i.e., η, the number of neutrons liberated per neutron absorbed by the fuel. However, there are other possible ends to a neutron, even a thermal neutron; and a fission-produced neutron is far from thermal — there are many hazards twixt one fission event and the next. So let us follow what happens when n thermal neutrons are absorbed by the fuel. For the sake of argument, let us assume that the fuel is low-enrichment uranium. The concepts will be more generally applicable.

First, the η of Appendix C was only defined for a pure material. We want to apply it to whatever we call "fuel." This will include fissile and fertile material, and may include other constituents such as the cladding. If ν is the number of neutrons liberated per fission, η is, then, just ν multiplied by the ratio of the probability of fission to the total probability of neutron absorption in the fuel. If we specify the first probability by $N_{fissile}\sigma_f$, the second is

$$\sum_i N_i \sigma_{ia} \, ,$$

where N_i is the number density for the ith constituent of the fuel and σ_{ia} is the corresponding cross-section for neutron absorption. (See Appendix A for discussion of cross-sections.) We calculate η for the thermal neutrons available and obtain

$$\eta = \nu \frac{N_{fissile}\,\sigma_f}{\sum_i N_i \sigma_{ia}} \, .$$

And the absorption of n neutrons by the fuel yields $n\eta$ fission neutrons.

Before the $n\eta$ fast neutrons can be thermalized, some of them will be absorbed by the fuel, in which case they may induce fission, thus increasing the number of neutrons. Most of such events will be induced in ^{238}U, provided the neutron energy is still above ^{238}U's fission threshold. The number of neutrons slowing down past this threshold is then $n\eta\epsilon$ where ϵ is the "fast-fission factor," the meaning of which is obvious. The number η depends on the fuel, but ϵ depends on all aspects of the core. Its definition is more operational than fundamental. (It has a value within a few percent of 1 for thermal reactors.)

In continuing to slow down to thermal energies, some neutrons will be absorbed (without, for the most part, causing fission), particularly by sharp resonances. (An example of such resonances, but for fission rather than absorption, is given in Figure B-1.) The portion of neutrons escaping such an end is p, the "resonance escape probability," so that the total number of neutrons reaching thermal energies is $n\eta\epsilon p$. Of these, only a fraction, f, will be absorbed by the fuel, so that whereas n thermal neutrons of the previous generation were absorbed by fuel — $n\eta\epsilon pf$ of the present generation are so absorbed. Therefore, the multiplication factor $k = \eta\epsilon pf$. (We have avoided the question of finite reactor size and consequent leakage of neutrons from the system. To be precise, we would specify f to be the ratio of thermal neutrons absorbed in the fuel to those absorbed by all reactor materials, and $\eta\epsilon pf$ would really be the multiplication factor for an infinite system. We would then obtain k for a finite system by multiplying $\eta\epsilon pf$ by a "nonleakage probability.")

The resonance escape probability is a very complicated quantity, and it is, in fact, temperature dependent. One such dependency is called the "Doppler effect," a name suggested by the fact that the fuel nuclei are in thermal motion; these thermal energies clearly depend on the temperature. Consider the effect of an absorption resonance on the spectrum of neutron energies. The resonance robs neutrons from an energy region comparable in width to that of the resonance. If the neutron resonance is broad, the number of neutrons absorbed by the resonance could be obtained by integrating (over the resonance) the product of neutron flux[1] times absorption cross-section, where we could use for the flux the expected value in the absence of the resonance. We can do this because the flux at any given value of energy is only slightly modified if the resonance is broad. If, however, we sharpen the resonance considerably, but let the resonance keep the same strength, i.e., integral of the cross-section, the resonance will significantly decrease the flux at the resonance energy, so that *fewer neutrons will be absorbed,* since the integral of flux x cross-section will be less. Thus a very sharp resonance *masks itself.* There are many very sharp resonances through which the neutrons must pass. The thermal motion of the fuel nuclei, however, *broadens* the resonances, decreasing this masking effect, and thus effectively decreasing the total number of neutrons. An increase in temperature, therefore, *decreases k,* assuming that the temperature change does not affect the fission rate via other mechanisms. Hence the Doppler effect may constitute a negative "reactivity insertion" with increasing temperature (usually measured in terms of a "Doppler coefficient"), and is an important intrinsic safety characteristic, particularly for fast reactors.

Loss of coolant in a water-cooled reactor also decreases k because the neutrons are less effectively moderated. A void in the coolant of the fast reactor may *increase* k. Such reactivity insertions are to be avoided. Often, reactivity is measured in terms of "dollars" where one dollar corresponds to β, the delayed-neutron fraction. For example, individual control rods are often "worth" slightly less than one dollar, indicating that a single control rod could not be responsible for a catastrophic reactivity insertion. The effects of consuming the fissile fuel, producing or burning neutron poisons, changing the coolant density, etc., can all add or subtract reactivity and, therefore, are liable to measurement in dollars.

One of the most important fission product poisons in thermal reactors is ^{135}Xe, formed from the beta decay of ^{135}I; ^{135}Xe is a very strong neutron absorber, so that in a strong neutron flux it is destroyed primarily by neutrons. At any specified reactor power level, the amount of ^{135}Xe will be established by a balance between xenon production and its removal by neutron absorption and by beta decay of the xenon. If the power level is decreased suddenly, ^{135}Xe will continue to be formed by beta decay of its parent, and its concentration will increase for a time because of decreased neutron absorption. If an attempt is then made to bring the reactor back up to power, the effect of elevated levels of ^{135}Xe on the neutron population will have to be overcome. Until the previous balance is reestablished, excessive numbers of neutrons will be lost to xenon, so that, to keep a multiplication factor of 1, excess reactivity will have to have been built into the reactor. For "excess reactivity" read "extra fissile material." For normal reactor operation, this excess reactivity must be masked by control, one more reason for its incorporation into the reactor system. One alternative is to wait out the xenon transient, i.e., wait until the ^{135}Xe decays away; this can result in a substantial shutdown period, because ^{135}I and ^{135}Xe have half-lives of 7 and 9 hours, respectively, and, therefore, the concentration of xenon increases rather slowly before finally decreasing. Another alternative is to build in only enough excess reactivity to overcome xenon provided the reactor is returned to power soon after the power decrease, and thus before xenon has grown in very much.

The isotope ^{135}Xe is also a noticeable steady-state poison; at typical neutron fluxes, it may absorb a few percent as many neutrons as are absorbed by ^{235}U. This is a significant portion of the total absorption by all the fission products.

---

[1] The neutron flux at a point in space is basically the product of the number of neutrons (of the energy being considered) per unit volume and the speed of the neutrons. If all these neutrons were moving in the same direction, the result is simply the rate at which neutrons pass through a unit area perpendicular to this direction; this is consistent with the ordinary usage of the word "flux."

APPENDIX F

The Nuclear Fuel Cycle

Production of nuclear fuel and its disposition after use involve a number of important steps which, together with the power plant itself, constitute the nuclear fuel cycle. Many aspects of this fuel cycle were treated in Chapter 11, but it is useful to indicate in somewhat greater detail some of the important aspects of nuclear fuel cycles. Of particular interest at the present time is the light-water reactor fuel cycle, both because of the LWR's prevalence in the United States and elsewhere and because this fuel cycle would tie in closely with that of the liquid metal fast breeder reactor, which is maintaining its prominence as the reactor of the future in most parts of the world.

The major facilities of the LWR fuel cycle are indicated in Figure 11-2. The processes at the "front" end of the fuel cycle, from uranium acquisition to fuel fabrication, are reasonably well established on a production basis, a necessary feature of the commercialization of nuclear power. Establishment of many of these facilities, from uranium mines through enrichment plants, was required by military programs. With the sale of numerous reactor power plants to utilities, both reactor vendors and other businesses have established commercial fuel fabrication plants, aided by experience with making fuel for military and research reactors.

The incentives for operating other parts of the fuel cycle have not been as overwhelming as the need for producing fuel. However, as discussed in Chapter 11, reprocessing of irradiated fuel to extract fissile material and to transform radioactive wastes into a form suitable for disposal maintains great interest. For many advanced fuel cycles, associated with some thermal reactors and particularly with fast breeder reactors, reprocessing is essential.

Some experience has been gained from reprocessing of fuels from AEC and ERDA (now Department of Energy) programs and from operation, for a time, of the Nuclear Fuel Services plant in New York. However, no commercial reprocessing plant is at present operating, at least not in the United States. However, a large-capacity separations facility is now ready for operation at Barnwell, South Carolina; as discussed in Chapter 11, the licensing of this facility must await resolution of questions about the form in which various products are to be converted, as well as more fundamental questions about fuel recycle in general.

The fuel cycle that we describe in this Appendix is for the LWR system. In many respects, fuel cycles for other systems would be similar except for rather obvious modifications (see the end of this Appendix).

LIGHT-WATER REACTOR FUEL CYCLE OPERATIONS

The highest grade uranium deposits in the United States occur primarily in the Colorado plateau, including parts of Arizona, Colorado, New Mexico, and Utah, as well as in Wyoming. Ores from these deposits usually contain less than 1% uranium, so that substantial processing is required, largely in a mill near the mine site, to obtain any relatively pure compound of uranium. Although it is possible to achieve some separation of uranium by physical methods, the basic scheme is to crush the ore, then to leach out the uranium by mixing the crushed ore with an acidic or alkaline solution (sulphuric acid is often used). The uranium that goes into solution is then extracted by a sequence of chemical operations (such as ion exchange or solvent extraction, followed by precipitation and drying) which leave it in the form of U_3O_8 concentrate with a purity of about 70%. This product of the milling operation is sealed in drums for transport to the UF_6 conversion plant. An important remnant of the milling operation is a "tailings" pile (see below under "environmental releases").

It is next necessary to purify the U_3O_8 and produce uranium hexafluoride (UF_6), the compound that is used in the enrichment plant. It is particularly important to remove any impurities that would reduce the quality of the uranium as a nuclear fuel, especially those isotopes with large neutron-absorption cross-sections. The purification is achieved by an extraction process, the product of which is oxidized to UO_3. This is converted by treatment with hydrogen, hydrogen fluoride, and fluorine to UF_6, a solid at room temperatures, which is shipped to gaseous diffusion plants for enrichment.

At the diffusion plant, uranium is processed to increase the concentration of ^{235}U above the 0.7% content of natural uranium. Use is made of the fact that UF_6 containing ^{235}U has a lower molecular weight than that containing ^{238}U. At thermal equilibrium, the lighter molecules will have the same average energy as the heavier, and, therefore, will be somewhat faster. When heated to a gaseous form, the lighter molecules, because of their greater speed, strike any containing walls more often, and therefore diffuse through a porous barrier slightly faster than the heavier (see Figure 11-12). The enrichment plant consists of a large number of such barriers in a "cascade" where, at a given stage, the enriched product proceeds to the next higher stage, while the depleted remnant recycles to the next lower. In principle, a cascade can be designed long enough to yield a product arbitrarily pure in ^{235}U. In practice, the highest commercial grade of uranium is 93% ^{235}U, which would be used in the HTGR; LWRs require 2% to 4% enrichment. The waste stream ("tails") from the enrichment plant typically has a ^{235}U content of 0.2% to 0.3%.

The UF_6 with the desired enrichment is shipped to fuel fabrication plants. A typical process line would then expose the UF_6 to ammonium hydroxide, yielding ammonium diurynate which is dried, then reduced (using hydrogen) to UO_2. This is made into a powder which (for water-cooled reactors) is pressed into a pellet, then sintered (heated) to form a ceramic. This pellet is ground, finally, to cylindrical shape, then incorporated into fuel pins and assemblies, as discussed in Part II.

In LWRs, fuel remains in the reactor for about three years of operation, a portion of the fuel being replaced each year. On removal, the fuel is stored in water pools (in the power plant building), which provide cooling and shielding, for several months; during this time, the fuel's radioactivity decreases substantially, due to

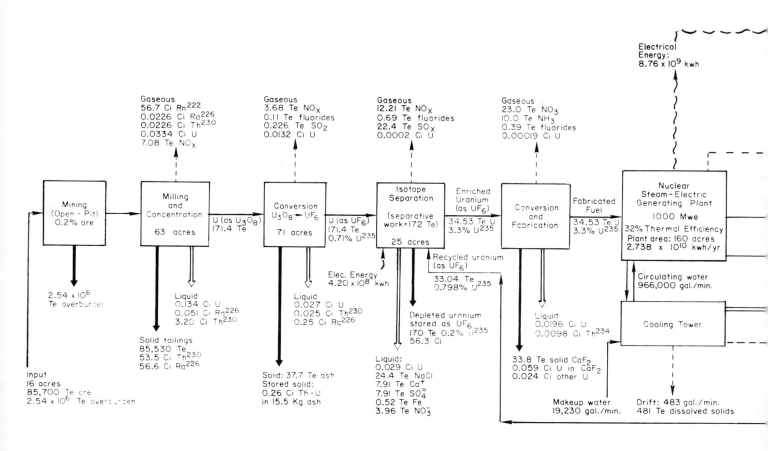

decay of short-lived species. The fuel, cooled and shielded by special shipping casks, would then be transferred to a fuel reprocessing plant, were such now available.

At the reprocessing plant, hardware is removed from the assemblies, the pins are chopped up and placed in nitric acid, which dissolves the fuel. (This is often described as the "chop-leach" method.) The remaining cladding hulls are relegated to waste storage. Uranium and plutonium are removed from this acidic solution by solvent extraction, leaving the high-level wastes, consisting primarily of fission products (in acidic solution). The plutonium may then be separated by another solvent extraction, and the uranium purified by processes similar to those at the UF_6 conversion plant (appropriate, since one of the products most desired would be UF_6 for return to the enrichment plant). A flow diagram for an LWR fuel reprocessing plant is shown in Figure 11-3. The separations facility of a full-scale commercial reprocessing plant is now complete at Barnwell, S.C. This plant has a throughput of 1500 metric tons of heavy metal (uranium and products) per year, equal to the output of about 50 LWRs. A major question in licensing of this plant will be uncertainty in the handling of plutonium, particularly to prevent its diversion or illegal acquisition. In a complete LWR cycle, the uranium would be returned to enrichment plants and the plutonium to fuel fabrication facilities that are designed to permit handling of plutonium in a way that adequately protects plant workers.

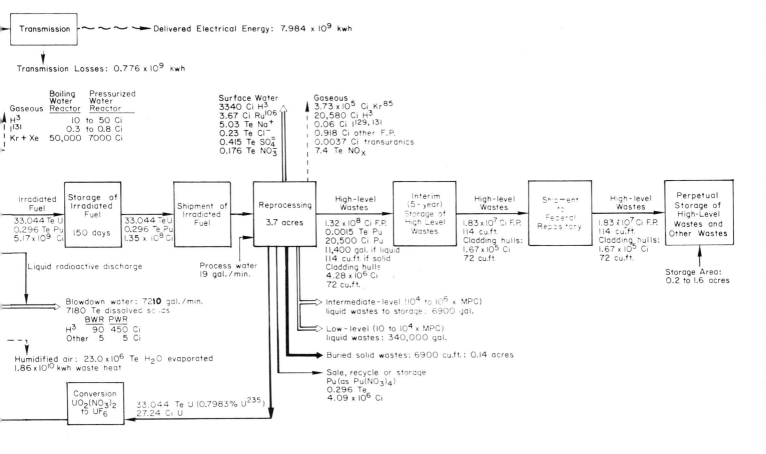

Figure F-1. LIGHT-WATER REACTOR SYSTEM WITH URANIUM FUEL: MATERIAL AND ENVIRONMENTAL RELEASE FLOWSHEET.
Material flows and environmental releases are shown for a uranium-fueled LWR fuel cycle in which fuel is reprocessed and uranium is recycled, but plutonium is stored. The values are for a 1000 MWe plant operated at 100% capacity factor. (Figure reproduced from Pigford et al., EPA report PB-259-876).

The high-level liquid wastes are an acidic, highly radioactive solution which can be consolidated somewhat by evaporation, but ultimately must be reduced to a solid. Methods are available for "calcining" these wastes to a dry, granular solid which can be stored on an interim basis in stainless steel bins or capsules. But the ultimate disposal that is most favored is a fission-product-bearing glass which would be deposited in geologic formations, as discussed in Chapter 11. The details of this are not settled, nor has there been any need due to the lack of any commercial reprocessing.

Figure F-1 is a material and environmental flowsheet for a light-water reactor system that includes reprocessing and recycling of uranium, but not plutonium. Figure F-2 is a comparable flowsheet for a light-water reactor that is fueled with natural uranium and plutonium from its own spent fuel supplemented by plutonium from about two uranium-fueled LWRs: for each of *three* reactors, only two would require enriched uranium; the third would operate on natural uranium plus

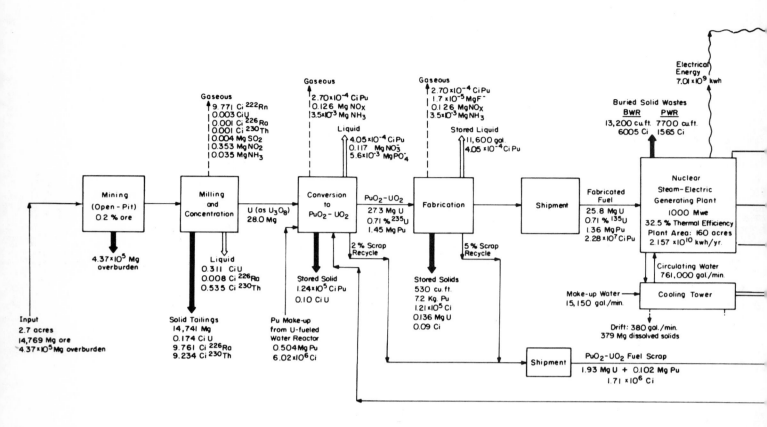

Gaseous
↑9.771 Ci ^{222}Rn
| 0.003 Ci U
| 0.001 Ci ^{226}Ra
| 0.001 Ci ^{230}Th
| 0.004 Mg SO$_2$
| 0.353 Mg NO$_2$
| 0.035 MgNH$_3$

Gaseous
↑ 2.70×10^{-4} Ci Pu
| 0.126 Mg NO$_x$
| 3.5×10^{-3} Mg NH$_3$

Liquid
↑ 4.05×10^{-4} Ci Pu
| 0.117 Mg NO$_3$
| 5.6×10^{-3} MgPO$_4^-$

Gaseous
↑ 2.70×10^{-4} Ci Pu
| 1.7 ×10^{-5} MgF$^-$
| 0.126 Mg NO$_x$
| 3.5×10^{-3} Mg NH$_3$

Stored Liquid
↑ 11,600 gal.
4.05 ×10^{-4} Ci Pu

Electrical
Energy
7.01 ×10^9 kwh

Buried Solid Wastes
BWR   PWR
13,200 cu.ft. 7700 cu.ft.
6005 Ci 1565 Ci

Mining (Open-Pit) 0.2 % ore
4.37×10^5 Mg overburden

Milling and Concentration
U (as U$_3$O$_8$) 28.0 Mg

Conversion to PuO$_2$-UO$_2$
PuO$_2$-UO$_2$ 27.3 Mg U 0.71 % ^{235}U 1.45 Mg Pu

Fabrication
2 % Scrap Recycle
5 % Scrap Recycle

Shipment
Fabricated Fuel 25.8 Mg U 0.71 % ^{135}U 1.36 Mg Pu 2.28 ×10^7 Ci Pu

Nuclear Steam-Electric Generating Plant 1000 Mwe 32.5 % Thermal Efficiency Plant Area: 160 acres 2.157 ×10^{10} kwh/yr.

Circulating Water 761,000 gal./min.

Make-up Water 15,150 gal./min.

Cooling Tower

Drift: 380 gal./min. 379 Mg dissolved solids

Input
2.7 acres
14,769 Mg ore
4.37×10^5 Mg overburden

Liquid
0.311 Ci U
0.008 Ci ^{226}Ra
0.535 Ci ^{230}Th

Solid Tailings
14,741 Mg
0.174 Ci U
9.761 Ci ^{226}Ra
9.234 Ci ^{230}Th

Stored Solid
1.24×10^5 Ci Pu
0.10 Ci U

Pu Make-up from U-fueled Water Reactor
0.504 Mg Pu
6.02×10^6 Ci

Stored Solids
530 cu. ft.
7.2 Kg. Pu
1.21×10^5 Ci
0.136 Mg U
0.09 Ci

Shipment
PuO$_2$-UO$_2$ Fuel Scrap
1.93 Mg U + 0.102 Mg Pu
1.71 ×10^6 Ci

4.37×10^5 Mg overburden

plutonium from all three. (An alternative plutonium recycle scheme is "self-generated" recycle, where each reactor uses only the plutonium which it generates, thereby reducing the required uranium enrichment.)

ENVIRONMENTAL RELEASES

The most significant releases of radioactivity from routine fuel cycle operations occur at the mining and milling site, the nuclear power plant itself, and the fuel reprocessing plant. The most important source of accidental releases is the nuclear power plant.

The routine releases from the mine site have a different origin than from other facilities, being due to daughters of ^{238}U already present in the ore when it was mined. In particular, radium 226 — to which ^{238}U decays in several steps — decays by a alpha emission to radon 222, a gas that becomes available to human ingestion (in spite of the fact that it is noble) because it decays to ^{218}Po. This substance is so chemically active that it immediately attaches itself to particulates in the air, assuming that the radon has escaped to the atmosphere. Daughters of ^{218}Po may then remain attached. In any event, they are available for collection by human respiratory membranes on breathing, and the various decays in the sequence can thereby expose humans to radiation. Radon is also one component of natural radioactive background; the population group that has shown the effect (high lung

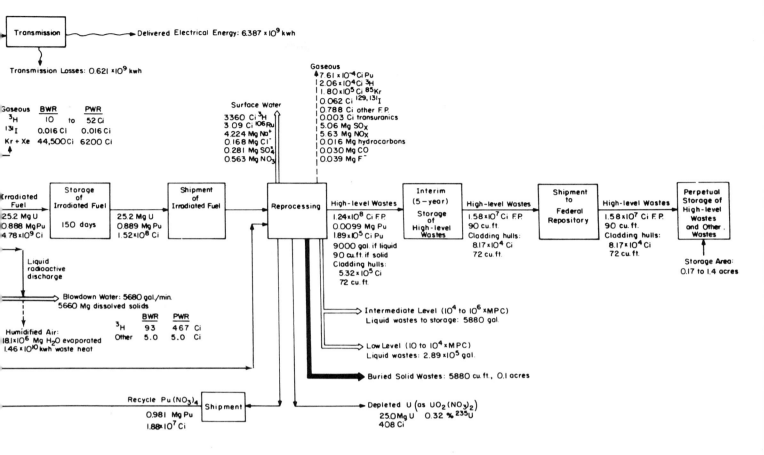

Figure F-2. LIGHT-WATER REACTOR SYSTEM WITH NATURAL URANIUM AND RECYCLED PLUTONIUM FUEL: MATERIAL AND ENVIRONMENTAL RELEASE FLOWSHEET.

Material flows and environmental releases are shown for an LWR system in which the LWR is fueled with natural uranium plus plutonium from its own spent fuel and from fuel of uranium-fueled LWRs. The values are for a 1000 MWe LWR operated at 80% capacity factor. (Figure reproduced from Pigford et al., EPA report PB-259-876.)

cancer rates) of breathing elevated levels of radon and its daughters are uranium miners. Other groups may also suffer from elevated levels because of the presence of the tailings pile at the mill site. This makes a substantial quantity of ^{238}U-daughters available to the open atmosphere, a fact that is significant when tailings have been used for landfill under homes and buildings, or for other purposes.

Radioactivity released from power plant and reprocessing sites is generally produced in the course of the nuclear chain reaction or subsequently. The most significant routine releases by far are the gaseous species tritium (^{3}H), produced by fission or formed by neutron capture on deuterium, carbon 14, produced from neutron reactions with nitrogen and oxygen, and krypton (^{85}Kr) or xenon (^{135}Xe), noble gas fission products. The tritium becomes incorporated in water, and is thus available for routine release from an LWR. Carbon 14 can escape as carbon dioxide and other compounds, particularly at the reprocessing plant, unless steps are taken to retain it. The noble gases are to some extent released to the

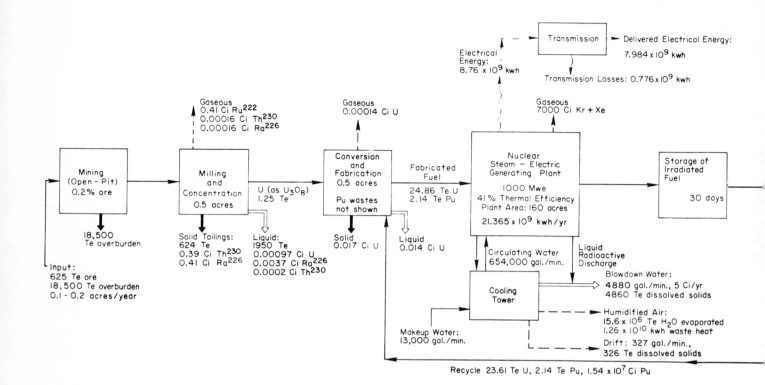

reactor coolant, from which they could routinely escape to the atmosphere, but more significantly, [85]Kr would ordinarily be released at the reprocessing plant unless the steps necessary to contain a noble gas were taken. The [85]Kr would cause the largest routine exposure to humans, presuming it were all released. (For a 1000-reactor system operating worldwide, it would be about 1 mrem/year skin dose, as compared with a natural background whole body dose in the vicinity of 100 mrem/year.) The carbon 14 has received much attention because, due to its long half-life (5730 years), it is a long-term environmental commitment.

It should be noted that all the routine releases mentioned are of gaseous products or isotopes (tritium) which are chemically identical to a major volatile constitutent of the reactor system. Accidental releases, however, can involve a broader range of materials simply because accident conditions may make other escape routes available. Barriers against release concentrate on two general classes of material: the fuel material itself (uranium and plutonium) and by-products of running nuclear reactors. (This is not an unambiguous distinction since plutonium and some isotopes of uranium are produced in reactors, and portions of the material produced would end up in the waste stream at the reprocessing plant.)

Fuel materials occur throughout the processes at the front end of the fuel cycle and continue to the reprocessing plant. For the plutonium recycle case, plutonium occurs from fabrication to reprocessing. (For a thorium-based cycle, [233]U would replace plutonium in this part of the cycle.) The fabrication process, the reactor, transportation, and reprocessing must all be regarded as possible sites for plutonium releases, and probabilities must be assigned to each release sequence.

Control of the large amount of radioactivity generated in the reactor is one

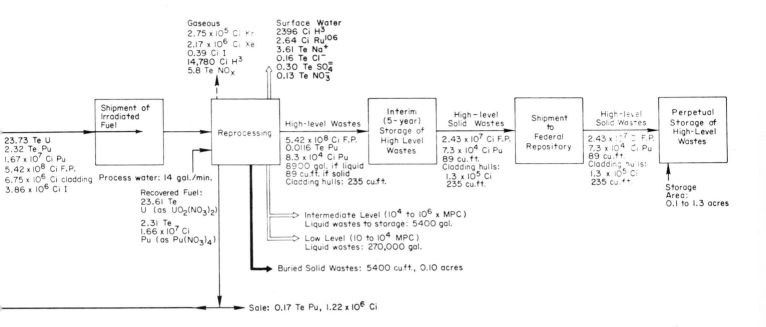

Figure F-3. LIQUID-METAL FAST BREEDER REACTOR SYSTEM: MATERIAL AND ENVIRONMENTAL RELEASE FLOWSHEET.
Material flows and environmental releases are shown for an LMFBR system. Both uranium and plutonium are recycled. Values are for a 1000 MWe plant operated at 100% capacity factor. (Figure reproduced from Pigford, et al., EPA report PB-259-876.)

purpose of the reprocessing and waste-management program. The most carefully analyzed sites for accidents making these materials available to the general environment are the reactor itself, a point where much energy is available to initiate releases, the fuel reprocessing plant, where the radioactivity is freed from the fuel assemblies, and the disposal site, where the wastes must be contained for long periods of time.

MODIFICATIONS FOR "ADVANCED" REACTOR SYSTEMS

Some obvious changes to the fuel cycle discussed in the previous section and indicated in Figures F-1 and F-2 would be necessary if converter reactors based on thorium or breeder reactors were to come into use. Many of these matters are discussed in Chapter 11. As one example, consider a converter where the initial feed is ^{235}U plus thorium as the fertile material. Eventually, most of the fissile feed is ^{233}U generated from thorium, but this uranium does not require enrichment (since it would not be "diluted" by ^{238}U). Thus ^{233}U effectively replaces plutonium in the fuel cycle flowsheet; but any ^{235}U that is supplied has to be highly enriched (although this might not be necessary in some schemes). Moreover, an auxiliary mining and milling branch must be established to supply thorium to the fabrication plant, and thorium recovered from reprocessing would be recycled (probably after

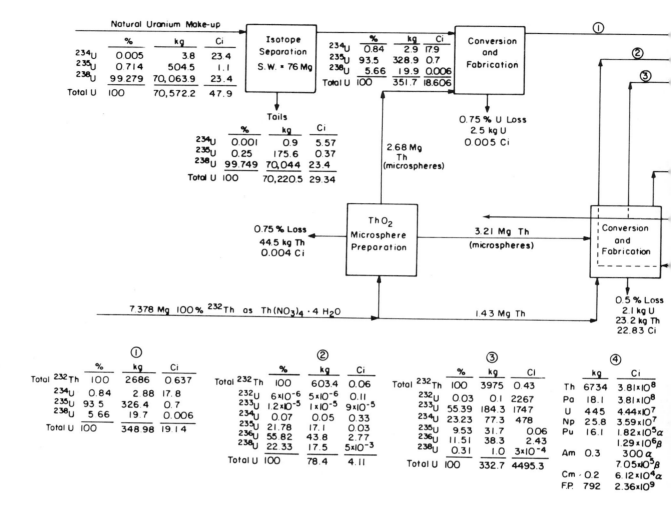

holdup to allow ^{228}Th to decay). Additional changes in handling of recycled material would be necessary because of penetrating radiation associated with a daughter of the ^{232}U that is produced in small amounts in the reactor.

A breeder reactor system as envisioned at present would differ from the LWR fuel cycle in one striking way: because the fissile material is primarily plutonium, no enrichment of uranium feed is necessary. In fact, the ^{235}U-depleted enrichment tails would be used. Effectively, then, all of the breeder fuel would be LWR by-products or material recycled in the breeder fuel cycle.

Figure F-3 is a flowsheet for the LMFBR. As detailed information has not been prepared for thorium fuel cycles, Figure F-4 indicates basic material flows for a standard HTGR system.

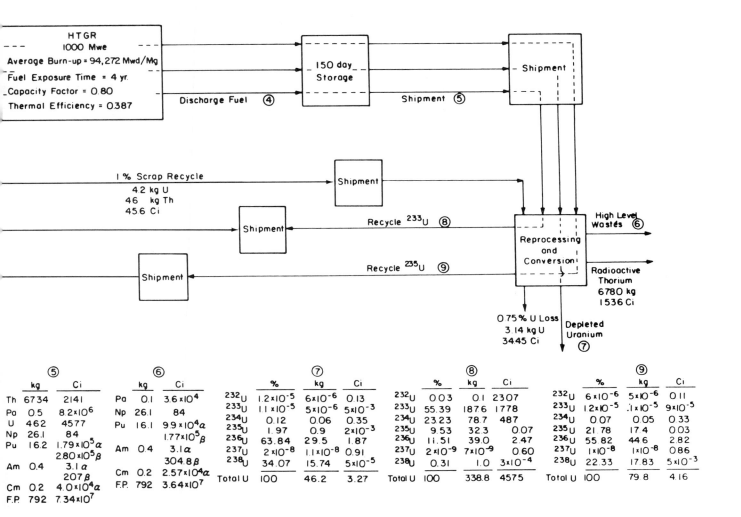

⑤			⑥				⑦				⑧				⑨		
	kg	Ci		kg	Ci		%	kg	Ci		%	kg	Ci		%	kg	Ci
Th	6734	2141	Pa	0.1	3.6×10⁴	^{232}U	1.2×10⁻⁵	6×10⁻⁶	0.13	^{232}U	0.03	0.1	2307	^{232}U	6×10⁻⁶	5×10⁻⁶	0.11
Pa	0.5	8.2×10⁶	Np	26.1	84	^{233}U	1.1×10⁻⁵	5×10⁻⁶	5×10⁻³	^{233}U	55.39	187.6	1778	^{233}U	1.2×10⁻⁵	1×10⁻⁵	9×10⁻⁵
U	462	4577	Pu	16.1	9.9×10⁴ α	^{234}U	0.12	0.06	0.35	^{234}U	23.23	78.7	487	^{234}U	0.07	0.05	0.33
Np	26.1	84			1.77×10⁵ β	^{235}U	1.97	0.9	2×10⁻³	^{235}U	9.53	32.3	0.07	^{235}U	21.78	17.4	0.03
Pu	16.2	1.79×10⁵ α	Am	0.4	3.1 α	^{236}U	63.84	29.5	1.87	^{236}U	11.51	39.0	2.47	^{236}U	55.82	44.6	2.82
		2.80×10⁵ β			304.8 β	^{237}U	2×10⁻⁸	1.1×10⁻⁸	0.91	^{237}U	2×10⁻⁹	7×10⁻⁹	0.60	^{237}U	1×10⁻⁸	1×10⁻⁸	0.86
Am	0.4	3.1 α	Cm	0.2	2.57×10⁴ α	^{238}U	34.07	15.74	5×10⁻⁵	^{238}U	0.31	1.0	3×10⁻⁴	^{238}U	22.33	17.83	5×10⁻³
		207 β	F.P.	792	3.64×10⁷	Total U	100	46.2	3.27	Total U	100	338.8	4575	Total U	100	79.8	4.16
Cm	0.2	4.0×10⁴ α															
F.P.	792	7.34×10⁷															

Figure F-4. HIGH-TEMPERATURE GAS-COOLED REACTOR SYSTEM FUEL
FLOWSHEET.
Material flows and isotopic compositions are shown for a 1000 MWe HTGR operated at 80%
capacity factor. (Figure reproduced from Pigford, et al., EPA report PB-259-876.)

Glossary

absorption, neutron — any reaction in which a free neutron is absorbed by a nucleus, including capture and fission

accelerator — a device that increases the speed, and hence energy, of particles, usually through forces induced by electromagnetic fields

accumulator (PWR) — a storage tank containing water under pressure and available for automatic injection into a PWR during a LOCA

actinides — a chemical group of heavy elements, including actinium, thorium, protactinium, uranium, neptunium, americium, and curium

alpha particle — a nuclear particle, which may be emitted during nuclear decay, consisting of two protons and two neutrons; the nucleus of ^{4}He

beta particle — an electron or positron (positively charged electron) emitted during nuclear beta decay

blanket — see *core*

blowdown — the rapid emptying of coolant from the primary system during a LOCA

break-even or *self-sustaining* — in connection with reactor neutronics, the condition with conversion ratio equal to 1, i.e., where as much fissile material is produced as is destroyed

breeding — production of fissile material (usually in excess of that consumed) from fertile material

breeding ratio — the ratio of the number of fissile atoms produced to the number consumed (usually of the same kind)

burnup — the percentage of heavy metal (i.e., fuel) atoms fissioned *or* the thermal energy produced per mass of fuel

capacity factor — ratio of average plant electrical energy output to rated output

capture, neutron — a reaction in which a nucleus absorbs a neutron (and may emit a gamma ray, but does not fission)

centrifuge — a device that separates materials of different density by rotating them rapidly, thereby subjecting them to "centrifugal" forces

chain reaction, nuclear — the sequence of reactions in which neutrons, the products of fission reactions, induce subsequent fission reactions

cladding — the material that surrounds the nuclear fuel material; in many reactors, the cladding is a metal can

275

condenser — heat exchanger in which steam is transformed into (liquid) water by removing heat (and transferring it to a cooling river, pond, etc.)

containment — a structure designed to contain the products (primarily radioactive) of any abnormality

control — neutron absorbing mechanisms used for maintenance of the multiplication factor (and hence of the chain reaction) at the desired level; in a more general sense, may refer to any systems (including electronic instrumentation) for maintaining operational direction of the power plant

control rods — rods containing neutron absorbing material; an important form of control for any commercial reactor

conversion — process whereby fertile material is changed to fissile material; may also refer to transformation of nuclear fuel or waste materials from one chemical form to another

conversion ratio — the ratio of fissile atoms produced to fissile consumed

coolant — the fluid that removes the nuclear generated heat from the core; in an LWR, the coolant is also the moderator

core — the region of a reactor containing the nuclear fuel. It is in this region that the nuclear reactions occur, with the exception of those caused outside the core by escaping radiation or radioactivity. Although the "core" is usually used to refer to all the fuel assemblies, it may also refer specifically, in the case of breeders, to those assemblies with substantial fissile loading. However, this may also be called the "seed" region. In either case, this region may be distinguished from the "blanket," which is intended primarily to contain fertile material in which fissile material can be generated by neutrons escaping from the core or seed region

critical mass — the minimum mass for a barely critical chain reacting system of specified composition, assuming criticality is possible for that composition

critical size — the minimum size to just yield criticality for a specified core composition and shape, assuming criticality is possible for that composition

criticality — a condition where the number of neutrons (or fissions) in a chain reacting system is the same from one generation to the next

cross-section — a measure of the probability of interaction between two particles for specified initial conditions and results

decay, radioactive — the process by which a nucleus of one type transforms into another, accompanied by emission of radiation

decay heat — the heat produced by radioactive decay of materials that are primarily the remnants of the chain reaction

delayed neutrons — neutrons emitted after radioactive decay, and hence which appear later (thereby making control easier) than the reaction that produced the decaying nucleus; the fraction of neutrons that is delayed is given by β.

denaturing — in connection with nuclear proliferation, reducing the usefulness of fissile material for weapons by diluting it isotopically

deplete — to reduce the fissile content of an isotopic mixture, particularly uranium

diffusion — the process in which molecules move or mix through random thermal motion

dose — the amount of radiation received by an organism as measured in energy deposited per mass of tissue (unit is the rad)

dose equivalent — the radiation dose multiplied by the relative biological effectiveness for the radiation type (unit is the rem)

doubling time — the time during which a reactor system produces excess fissile material equal to the system inventory

electron — a negatively charged atomic particle whose mass is 1/1836th that of a proton *emergency core cooling system* — any engineered system for cooling the core in the event of failure of the basic cooling system; may include core sprays, injectors, etc.

enrichment — the percentage of fuel atoms that are fissile; the process of elevating the percentage of a particular isotope in an isotopic mixture; usually the isotope of interest is ^{235}U

eta (η) — the number of neutrons produced per neutron absorbed in a specified fuel material

event tree — an analytical device for identifying the possible sequences that may result from choices and for calculating the probabilities associated with each sequence

exclusion area — the area immediately surrounding a nuclear facility and to which the public does not have free access; often is just the facility site

exposure — see *dose, dose equivalent*

fast neutrons — neutrons that have not been thermalized, i.e., having energy comparable to their energy immediately after production from fission

fast reactor — a reactor that does not moderate fission neutrons and that is intended to take advantage of the higher neutron yield from fission induced by fast neutrons

fault tree — an analytical device in which probability and modes of system failure are determined by identifying how subsidiary systems contribute to failure; contributors to subsystem failure are then identified, and so on, until reaching components whose failure rates are known

feedwater — water, usually from a condenser, supplied to replenish the water inventory of any component, but particularly a boiler or steam generator

fertile nucleus — a nucleus that may capture a neutron to form a product that eventually decays to become a fissile nucleus

fissile — capable of being split by interaction with a thermal neutron

fissile loading — the percentage of fuel atoms in a reactor that are initially fissile; or the total initial mass of fissile material in the reactor

fissile nucleus — a nucleus that can be induced to fission by a "slow" (thermal) neutron

fission — the splitting of a heavy nucleus to form two lighter "fission fragments," as well as less massive particles, such as neutrons

fission products (fragments) — the medium-weight nuclear products (ordinarily two per fission) resulting from the splitting of a heavy nucleus

fissionable nucleus — a nucleus that can be induced to fission by a neutron

fuel — basic chain-reacting material, including both fissile and fertile materials

fuel pellet, rod (pin), cladding (coating), assembly (bundle) — basic form in which fuel is present in solid-fueled reactors

fusion — the process of joining light nuclei, via nuclear reactions, to form heavier nuclei; often refers to reactions among hydrogen isotopes, deuterium and tritium

gamma rays — high-energy electromagnetic radiation emitted from nuclei as a result of nuclear reactions or decay

half-life — the time during which half of the nuclei in a sample of a particular radioactive nuclide will decay

heat rate — inverse of plant efficiency, usually given in units of Btu/kWh

heavy metal — in connection with fission reactors, the fuel materials uranium, plutonium, thorium, and their reaction products

"high gain" — refers loosely to reactors with high-conversion ratios

hot (cold) leg — the portion of a coolant circuit (usually primary) through which the coolant exits (enters) the site at which it is heated, i.e., the reactor vessel

Inconel — an alloy, principally of nickel, iron, and chromium, typically used for the tubing in a PWR steam generator

initial uranium requirement — the amount of natural uranium (usually measured as tons of U_3O_8) required to produce the initial fuel load for a reactor that is beginning operation

inventory, fuel — the fuel (or amount thereof) contained in a reactor and, possibly, in associated storage, reprocessing, etc.

ionization — the process of creating oppositely charged pairs of particles; often refers specifically to removal of electrons from atoms or molecules

isotope — a particular species of a given element; the element is specified by the proton number, and the isotope by mass (proton plus neutron) number

laser — a device that produces a coherent, and usually intense and well-collimated, beam of electromagnetic radiation of well-determined wavelength or energy by "light amplification by stimulated emission of radiation."

lifetime uranium requirement — the total amount of natural uranium (often in tons of U_3O_8) necessary to supply fuel for a reactor's lifetime, including both initial loading and refueling

linear hypothesis (dose-response) — the hypothesis that latent responses (such as cancer) to radiation exposure depend linearly on the dose equivalent

linear power — usually refers to power generation per length of fuel rod

loss-of-coolant accident (LOCA) — a reactor accident in which coolant is lost from the primary system

low population zone — a zone around a nuclear facility in which the population is small enough for evacuation to be effective in a short enough time to avoid large radiation doses, should a large radioactive release occur

makeup uranium requirement — the amount of natural uranium (typically in tons of U_3O_8) necessary to produce fresh fuel for routine refueling of a reactor

maximum permissible concentration — the maximum concentration of specific radionuclides in environmental media (air and water), determined by calculation of the dose equivalent resulting from exposure of workers (for 40 hours per week) or of the public (for 168 hours per week)

moderator — any material chosen to slow neutrons by elastic collisions

multiplication factor — the ratio of neutron (or fission) numbers in succeeding generations of a chain reaction

net energy — the available energy from a system, i.e., the energy produced by the system, less the energy to produce and operate the system

neutron — an uncharged nuclear particle, one of the two principal components of nuclei, with mass similar to that of a proton

neutron poison — any material excepting fuel materials, which absorbs neutrons; some poisons are used for control

nuclear steam supply system (NSSS) — the basic reactor and support equipment, plus any associated equipment necessary to produce the steam that drives the turbines

nuclide — a single nuclear type, specified by proton and neutron numbers (the sum of which is the mass number)

once-through fuel cycle — a nuclear system wherein nuclear materials are introduced into a reactor only once; they are not recycled

plant efficiency — see *thermal efficiency*

plasma — an ionized gas of positive ions and electrons

population dose — the summed radiation dose-equivalent to a population group (the unit is person-rem)

positron — an anti-electron having, therefore, positive charge

power density — the power generated per unit volume (usually of the core)

pressurizer — the device that maintains the pressure in the primary coolant circuit of a PWR within specified limits

primary coolant system (or loop) — the entire circuit through which the fluid that actually cools the core passes, including all piping, vessels, and components, such as the reactor vessel, coolant pump, steam generator, etc.

prompt neutrons — neutrons that appear immediately after the fission (or other) reaction responsible for their production

proton — a positively charged nuclear particle, one of the two principal components of nuclei, with mass similar to that of the neutron

pump (circulator) — a device that forcibly circulates a fluid, the most important example being the coolant

quality — the percentage, by weight, of the cooling water that is present as steam

radioactive — having the ability to decay spontaneously, thereby changing to another nuclide.

radionuclide — a radioactive nuclide

reaction, nuclear — an interaction between two (or more) nuclei, nuclear particles, or radiation, possibly causing transformation of nuclear type; includes, for example, fission, capture, elastic scattering

reactor — the core and its immediate container

reactor vessel — the container of the nuclear core or critical assembly; may be a steel pressure vessel, a prestressed concrete reactor vessel (PCRV), a low-pressure vessel (such as a calandria or sodium pot), etc.

recycle — the extraction of useful nuclear materials from irradiated fuel and their incorporation into fresh fuel

refill — the period after LWR loss of coolant during which injected water fills the reactor vessel to the bottom of the core

reflector — material placed around the core (and blanket, if any) to reflect escaping neutrons back into the core; the reflector, therefore, decreases neutron loss and critical size

reflood — the period during an LWR LOCA (and after refill), when injected water covers the core; see also *refill*

reprocessing — the dissolution of irradiated fuel and separation into waste and fuel fractions; may also include conversion to particular chemical and physical forms

resonance — a sharp increase in the cross-section as a function of energy, for a particular reaction

resonance capture — the capture of neutrons at resonant energies (see *resonance*)

scram — the rapid shutdown, via introduction of neutron absorbers, of the chain reaction

separative work — a measure of the amount of processing at an enrichment plant; measured in units of mass, kg

slow neutrons — neutrons at approximately thermal energy

steam generator (boiler) — a heat exchanger in which steam is produced by heat that is transferred from the primary coolant, or even from a secondary fluid

tails — the remnant of physical or chemical processing to increase the content of a particular substance; applies to the depleted streams from uranium and thorium milling and from uranium enrichment

tails percentage (enrichment) — the percentage ^{235}U in enrichment plant tails (the depleted stream from the plant); typically 0.2% to 0.3%

thermal efficiency — in a power plant, the ratio of net electrical energy produced to total thermal energy released in the reactor or boiler

thermal energy — can refer either to heat or to the average energy of particles in thermal equilibrium with (i.e., having the temperature of) their surroundings; at room temperature, thermal energy is about 1/40 eV

thermal reactor — a fission reactor in which neutrons are moderated to thermal energy in order to increase reaction probabilities (cross-sections)

transuranics — elements with atomic number greater than that of uranium (92); see table of elements in Appendix A

turbogenerator — device in which steam (or some other gas) is allowed to expand, thereby doing work that drives an electric generator

waste disposal — the disposition of nuclear wastes, including fission products or actinides, at a site for long-term or permanent storage or burial

X-rays — intermediate energy electromagnetic radiation, typically emitted during atomic transitions

yellowcake — the form of the U_3O_8 product of uranium mining and milling operations

Zircaloy — an alloy of zirconium used as fuel rod cladding in water-cooled reactors because of its low thermal neutron cross-section and good heat transfer properties

Index

Designer	Al Burkhardt
Composition	Trend Western
Lithography	Murray Printing Co.
Binder	Murray Printing Co.

Text	IBM Composer Press Roman
Display	Linocomp Optima
Paper	50 lb. Finch Title 94
Binding	Holliston Roxite B 53548